Henry Ruttan

Ventilation and Warming of Buildings

Henry Ruttan

Ventilation and Warming of Buildings

ISBN/EAN: 9783337258481

Printed in Europe, USA, Canada, Australia, Japan

Cover: Foto ©berggeist007 / pixelio.de

More available books at **www.hansebooks.com**

OF

BUILDINGS.

ILLUSTRATED BY FIFTY-FOUR PLATES,

EXEMPLIFYING

THE EXHAUSTION PRINCIPLE.

TO WHICH IS ADDED

A COMPLETE DESCRIPTION AND ILLUSTRATION OF THE VENTILATION OF RAILWAY CARRIAGES, FOR BOTH WINTER AND SUMMER.

BY THE
HON. HENRY RUTTAN,
LATELY VICE-PRESIDENT OF THE BOARD OF AGRICULTURE FOR UPPER CANADA, Etc., Etc., Etc.

NEW-YORK:
G. P. PUTNAM, PUBLISHER, 532 BROADWAY.

1862.

ACKNOWLEDGMENTS.

I FIND it a pleasure, as well as a duty, to acknowledge my indebtedness to Mr. E. P. HANNAFORD, now of the Engineers' Department of the Grand Trunk Railway Company, for the principal part of my drawings, for the formation of which he had nothing to go by except my description of what I wished to represent. Considering that he had nothing to work upon, or by, in the shape of precedent, and during the two years in which we were getting them up, I was constantly interrupting him by alterations and improvements, these diagrams or plates may fairly be considered as entitled to a good share of the credit of the work—if credit there be in it at all.

In this respect also I acknowledge my obligations to Mr. STEWART and Mr. LOUGH, civil engineers, of Port Hope.

TO
HIS EXCELLENCY
The Right Honorable Lord Viscount Monck,
GOVERNOR-GENERAL IN AND OVER THE PROVINCES OF BRITISH NORTH-AMERICA,
ETC., ETC., ETC.,

THIS BOOK
IS,
By permission, respectfully dedicated by
HIS OBEDIENT, HUMBLE SERVANT,
HENRY RUTTAN.

CONTENTS.

CHAPTER I.
INTRODUCTORY, ... 1

CHAPTER II.
OUR SENSES, . . 4

CHAPTER III.
WHAT VENTILATION IS, ... 8

CHAPTER IV.
HEAT, ... 11

CHAPTER V.
THE AIR-WARMER, ... 16

CHAPTER VI.
AIR AND WATER, ... 19

CHAPTER VII.
CELLARS, ... 24

CHAPTER VIII.
WOOLEN CARPETS, ... 27

CHAPTER IX.
COLD FEET, ... 31

CHAPTER X.
DRY CLOSETS, ... 35

CHAPTER XI.
ARCHITECTS AND ARCHITECTURE, ... 37

EXPLANATION OF THE PLATES.

Plate I.—HOW TO LAY THE LOWER OR MAIN FLOOR OF A HOUSE, . . 42

Plate II.—HOW TO CONSTRUCT A CHIMNEY TO EXHAUST THE AIR EITHER ABOVE OR BELOW THE FLOOR AT PLEASURE; AND HOW TO WARM AND VENTILATE OLD BUILDINGS WHERE NO PROVISION FOR DRAWING THE AIR UNDER THE FLOOR EXISTS, ... 43

Plate III.—OPEN BASE, . . 45

Plates IV., V., VI.—FOUL-AIR-GATHERING DUCT, . 45

Plates VII., VIII., IX.—ELEVATION AND PLAN OF DWELLING, . . 49

Plates X., XI.—FOUNDATION OF A DWELLING-HOUSE, BY WHICH THE PRACTICAL WORKING OF THE VENTILATION MAY BE ILLUSTRATED, . . . 49
DRY CLOSETS, ATTACHMENT TO FOUL-AIR-SHAFT, . . . 54

Plates XII., XIII.—VENTILATION OF OLD BUILDINGS, . 56

Plates XIV., XV., XVI.—VENTILATION OF SCHOOL-HOUSES, . 62

Plates XVII. to XXV.—VENTILATION OF PUBLIC BUILDINGS—JAIL, 71

Plates XXVI. and XXX.—FOUL-AIR-SHAFT OR CHIMNEY, . . 76

Plates XXVII., XXVIII., XXIX., XXXVI.—VENTILATION OF A CHURCH, 80

Plates XXXI. and XXXII.—PLENUM FRESH-AIR DUCTS, 84

Plate XXXIII.—OPEN CORNICE AND SMOKE-PIPE REGISTER, . 86

Plate XXXIV., ... 87

LII. and LIII.—AIR-WARMERS AND AIR-WARMING STOVES, . 87

LIV. DO. . . . 88

VENTILATION OF RAILWAY-CARRIAGES, ... 90

PLATES FOR DO., ... 95

WINTER VENTILATION, ... 100

PLATES FOR DO., ... 103

PLATES, ... 107

PREFACE.

ALTHOUGH a pioneer upon ventilation upon the "exhaustion principle," I can not of course expect entire immunity from criticism. In extenuation, however, and in justification of myself, I am bound to say, that such has been the lengthened period of time which has elapsed—more than nineteen years—since I first engaged in my experiments upon the subject, and I have, in some minor points, been so closely followed up, that my first suggestions have, and may now, perhaps, properly be considered *old*. I have no doubt but that some may be surprised to hear me call them *my* inventions; yet if those who question their originality as claimed by me will but think back to the year 1843, they will fail to find at that period any traces of the points in question, originating any where except with me. However, I am not jealous or envious upon this subject. I am only too glad to find that the subject of ventilation has been progressing. I only demand that the credit upon the points to which I allude shall be given where it is due.

My object in putting out this book has been to put the rising generation, and especially the young builders, upon "the right track," so that before long the "million"—the poor as well as the rich—may avail themselves of the inestimable advantages of pure air within their dwellings, instead of the foul atmosphere in which they have been obliged to live; for it costs nothing except a little brain-work of the architect or builder.

I hope that my diagrams and drawings, as well as my instructions, may be understood. I have endeavored to adapt them to the comprehension of the unlearned apprentice, and hope that this may not be considered derogatory to those whose opportunities of an education have been of a higher order.

I am aware that in my endeavor to make every thing plain to the mechanic, however educated, upon whom so much depends, and who may be unaccustomed to the investigation of subjects of a scientific nature, I may fail to make the general principles of the subject fully understood; whilst, on the other hand, to those of scientific attainments, my treatment of the subject may be considered much too diffuse, and as, perhaps, containing a fulness of explanation and illustration approaching to redundancy. But this is a subject of universal concern, and may therefore justify an error on the right side. No human being can *afford* to be ignorant upon this matter. No person having the charge of children, whether as parent, guardian, or teacher, can excuse himself for allowing those under his charge to be totally ignorant of the subject of ventilation.

It is quite evident that those who have gone before me in the investigation of this subject, have never lived in a cold climate, or they would never have suggested the extraction of the air in winter from the top of the apartment. This has probably arisen from two fundamental and popular errors; one is, that that part of the air which lies next the ceiling being the warmest, must necessarily be the *foulest*. The other, that warm or hot air *naturally* goes upward, and therefore that the top of the room was the proper point to let this air out.

This is not the proper place to argue the points here raised; suffice it to say, that every room which has the least pretensions to being ventilated, in winter especially, has the *purest* air lying at the *top;* and secondly, the warmed air, under all circumstances, if not prevented, *naturally* falls *downward.*

HENRY RUTTAN.

Coburg, C. W., *6th August*, 1862.

VENTILATION AND WARMING.

CHAPTER I.

INTRODUCTORY.

After nineteen years of hard work, and the expenditure of as many thousand dollars, I venture to launch forth the result of my experiments in relation to the warming and ventilation of buildings, in the shape of a book.

Having read every thing that I could lay my hand upon, regarding the subject, and predicating my experiments upon the information thus obtained, I invariably, at the end of each experiment, found myself at fault, and just as far from accomplishing the object I had in view as ever, producing nothing more than had been obtained before me, until at last I gave it up.

Still revolving the subject in my mind, I was struck with the similarity of idea running through the principle which so many learned men had adopted. So eminently is this the case, that when a person has read one of these authors, he has, so far as regards the principle of a natural process of ventilation, read them all; and this is readily accounted for by the fact, that nearly all these men lived in the Old Country, where they have a comparatively mild climate, and where a system graduated to that climate was needed. Living as I did in a cold climate, my object was to work out a plan which would not only secure a change of air, but also insure *warmth*, and an equalized temperature.

Now, here was an entirely new field, which no one had trodden before me; the warming, as well as the ventilation of a building, in a cold climate, by *natural* means, and by one and the same process. So throwing aside my books, I resumed the experiments, and soon found that no natural process of warming and ventilation could be produced unless *natural* laws were obeyed, throughout; that we could not bend these to *our* wills, or in any one jot or tittle contravene them with impunity.

Volume upon volume has been written, theory upon theory has been started, diagram upon diagram has been published, to show the different operations of air under different circumstances; and experiments without number have been made on the two subjects of ventilation and warming, each writer and experimenter attempting to reconcile the general principles of philosophy to his own particular theory, instead of working out his theory by the unerring principles of philosophy. Any system, to be philosophical, must be universal in its application. Without this universality, it must fall to the ground; and when we hear of Dr. Reid's system, and Dr. Wyman's system, and a host of others who have written upon this subject, each advancing ideas perhaps differing from the others, no clearer proof need be adduced, that this great subject has never yet attained to the dignity of a system at all. It is a mere patching up of a piece of machinery, by the stray wheels and component parts of several other pieces of machinery, in order to produce a desired result, but which, if we may judge by the progress of ventilation thus far, has ended in such a want of harmony in its working, as to leave the whole subject, for all practical and useful purposes, very much in the same state that it was in the beginning.

The construction of an efficient system of warming and ventilation, requires that all the details pertaining to it should be reduced to one *harmonious whole*, which shall be applicable to every thing. If not good in all cases, it is good for nothing. It must be adaptable to the palace and the cottage, to the ship and to the railway-carriage, to the habitations of animals, as well as those of men, and in addition, it must be attainable by the poor as well as the rich.

Several reasons may be assigned why we have never yet, in the colder parts of America, attained to *the* system of ventilation; I say the system, because there can be but one. The fact is, that our experimenters, hitherto, have never made any distinction in the climates of different countries, but have derived their ideas from what has been said and done in Europe, which, however applicable they may have been in those countries, are in many cases worse than useless when applied to the northern parts of this continent.

Another reason may be found in the fact, that our architects, also copying from the books written in the milder countries of Europe, where little provision for either warming or ventilation was required, have never considered themselves responsible for either, and therefore have never turned their attention to the study of either.

A third reason, and one which has operated more powerfully against it than either of the preceding, is the ignorance which has hitherto prevailed as to its necessity. This, however, is now rapidly giving way before the light which has of late years been shed upon the subject; and I think a continuation of the same views will soon consummate its complete overthrow.

The chief thing now required to the general adoption of a system of warming and ventilation by proprietors of buildings, is persons who will carry it out. For this purpose, it is necessary that our architects and builders should learn how; and however presumptuous it may appear to these gentlemen, for me, as a non-professional man, to attempt to teach them, nevertheless, that is the object I have in view in publishing this book. We must have men such as we never had before—architects and builders for a cold climate. They must learn to provide the building with lungs, for unless it breathes, the inmates can not breathe. A man might as well trust to the pores of his body as channels through which to obtain the requisite supply of air, as for a builder or family to trust to cracks and crevices, or the occasional opening of a door, for purposes of ventilation.

A man may be what is called a good mechanic, he may form a piece of work after a model, and make what is called a "good job," by manual labor alone; but *he* will never become a master in his trade or calling, whose powers of *mind* have no part in the performance. He must look into the reason of things, he must exercise and accustom his mind to comprehend, to grasp the whole subject, of which every stroke he strikes, or every stone or brick he lays, is but an infinitesimally small part. It is such a person, and such only, who ought ever to be permitted to meddle with the ventilation of our buildings.

Knowing the distrust with which all new ideas are received, of course I can not expect entire immunity from those whose prejudices have, by long custom and habit, become so fixed as to be utterly ineradicable; but from the young architect and builder, and from citizens generally, I hope for a fair, honest, and candid trial, and my confidence that I shall receive it is strengthened by the multiplied indications of progress by which we are all surrounded. It is to this class, therefore, that I more especially address myself, and I have, in my explanations and instructions, so adapted my language and illustrations, that those who have had only the most limited opportunities may not be entirely at a loss in getting at my meaning.

CHAPTER II.

OUR SENSES.

It must seem to many an apprentice, brick-layer, and joiner, all of whom I hope will study this book, a strange proceeding for me to attempt to convert him to a new theory by first attacking his senses of hearing, seeing, feeling, and smelling, by the statement that they are all at fault, and that none of them are to be depended upon! Yet I can assure him that such is actually the case, and that in this inquiry he will invariably be led astray if he places any confidence in them, irrespective of the philosophical examination of the laws of nature.

Of all the means, says Dr. Lardner, of estimating physical effects, the most obvious, and those upon which mankind place the strongest confidence, are the senses. The eye, the ear, and the touch, are appealed to by the whole world as the unerring witnesses of the presence or absence, the qualities and degrees, of light and color, sound and heat. But these witnesses, when submitted to the scrutiny of reason, and cross-examined, so to speak, become involved in inexplicable confusion and contradiction, and speedily stand self-convicted of palpable falsehood. Not only are our organs of sensation not the best witnesses to which we can appeal for exact information of the qualities of the objects which surround us, but they are the most fallible guides which can be selected. Not only do they fail in declaring the qualities or degrees of the physical principles to which they are by nature severally adapted, but they often actually inform us of the presence of a quality which is absent, and of the absence of a quality which is present.

The organs of sense were never, in fact, designed by nature as instruments of scientific inquiry; and had they been so constituted, they would probably have been unfit for the ordinary purposes of life. It is well observed by Locke, "that an eye adapted to discover the intimate constitution of the atoms which form the hand of a clock, might be, from the very nature of its mechanism, incapable of informing its owner of the hour indicated by the same hand."

The term heat, in its ordinary acceptation, is used to express a feeling or sensation which is produced in us when we touch a hot body. We say that the heat of a body is more or less intense, according to the degree in which the feeling or sensation is produced in us. The touch by which we acquire the perception of heat, like the eye, ear, and other organs, is endowed with a sensibility confined within certain limits; and even within these, we do not possess any exact power of perceiving or measuring the degree or quality by which the sense is affected. If we take two heavy bodies in the hand, we shall in many cases be able to declare that one is heavier than the other; but in what degree, or how much, our senses fail to inform us.

If we look at two objects, differently illuminated, we shall, in the same way, be in some cases able to declare which is the more splendid, but the exact difference in the illumination we shall be unable to decide. It is the same with heat; if the temperature of two bodies be very different, the touch will sometimes inform us which is the hotter; but if they be nearly equal, we shall be unable to decide which has the greater and which the less temperature.

Feeling can never inform us of the quantity of heat which a body contains, much less the relative quantities contained in two bodies. Heat, in its latent state, can never be felt at all; for example, ice-cold water and ice appear to be of the same temperature, but the difference is considerable.

If we hold the hand in water which has a temperature of about 90°, after the agitation shall have ceased we become wholly insensible of its presence, and shall be unconscious that the hand is in contact with any body whatever. We shall of course be altogether unconscious of the temperature of the water. Having held both hands in this, let us remove the one to water of a temperature of 200°, and the other to water of a temperature of 32°. After holding the hands for some time in this manner, let them be both removed and again inserted in the water at 90°, immediately we shall become sensible of warmth in the one, and cold in the other. If, therefore, the touch be in this case taken as the evidence of temperature, the same water will be judged to be hot and cold at the same time.

If in the heat of summer we descend into a cave, we are sensible of cold; but if in winter, we have a sensation of warmth. Now, a thermometer suspended in the cave will always show the same temperature.

Thus we see that the sensation of heat depends as much upon the state of our own bodies as upon the several agencies which excite the sensation. If we step out of a warm bath into air, at the same temperature, we shall experience a sensation of coldness, because air, being a more rare and attenuated substance, a less number of its particles are in contact with the body.

If we step into a room of a high temperature, say 120°, the carpet will feel cool, and the tiles of the hearth or chimney-piece will be insupportably hot. If we enter a room of low temperature, say 32°, the reverse is the case; the carpet will feel warm and the tiles and chimney-piece cold, yet the temperature is the same. If we wrap a thermometer in a blanket, and lay another upon a piece of marble, in a room of any temperature, the indications will be the same in both cases, yet to the touch the two bodies will be very different—the one will feel cold and the other warm.

The air of a room has to our sight no color, yet we know it is blue. We know that the sea is green, yet there is not the slightest indication of color in a glass full of the water. If, as we have seen, we remove our hands from water at 200°, to that which is at a temperature of 100°, it feels cold, yet we know it is warm. If, therefore, the senses of seeing and feeling, and in fact all others, are so imperfect, or rather, I should say, such erring guides, is it not reasonable that in all matters relating to our health we should have recourse to sources of information other than those which we know may lead us astray?

Of all the organs of sense, that whose nervous mechanism appears to be most easily deadened by excessive action, is that of smelling. The most delightful odors can only be enjoyed occasionally, and for short intervals. The scent of the rose, or the still more delicate odor of the magnolia, can be but fleeting pleasures, and are destined only for occasional enjoyment. He who lives in a garden can not smell a rose, and the wood-cutter in the Southern forests is insensible to the odor of the magnolia.

Persons who indulge in the use of artificial scents, soon cease to be conscious of their presence, and can only stimulate their jaded organs by continually changing the objects of their enjoyment.

But every day's experience must convince the most careless observer how little dependence can be placed upon the sense of smell. We move into a new tenement, for instance, and we are at once sensible of a difference of smell; but in a very few days we become accustomed and perfectly insensible to it. We walk into a different apartment of the same building even, perhaps our own, and we are distinctly sensible of a peculiarity of odor, but which soon passes away if we remain in the room. We notice the loathing with which a person enters one of our prisons; in a few days, however, he ceases to complain. Pass from the pure air into an unventilated bedroom, which has been occupied the night previous by a lodger, and you at once become sensible of an almost intolerable odor, yet the person who occupied it was unable to detect any thing of the sort. It is just so in the case of badly ventilated houses; the inmates soon become so accustomed to the foul air they are continually breathing, that it is an almost hopeless task to convince them that it is any thing but the most pure. In short, such is the uncertainty of our organs, that delicate and refined as they may be, if our health depended solely upon their indications, they would be worse than useless to the human family.

It is indeed wisely ordered that our organs of sense should be constituted for active and practical use, rather than that they should, by the delicacy or grossness of their sensation, render us miserable. And it is especially so with the eye. It has already been observed that the eye, which was capable of discovering the atoms of which the hands of a clock are composed, would fail to inform us of the hour indicated by the same hand. It may be added, that a pair of telescopic eyes, which would discover the molecules and population of a distant planet, would ill requite the spectator for the loss of that rude power of vision necessary to guide his steps through the city he inhabits, and to recognize the friends who surround him.

But although no dependence can be placed upon the manifestations of our senses as to what may be good or evil, useful or injurious, yet the Almighty has endowed men with a mind, and a capacity to investigate, scientifically, all subjects connected with his physical existence; and this he is as much bound to do, as he is to investigate those laws which are placed before us as a guide in our moral existence.

If, therefore, by ordinary observation, we can not *see* the contaminations of the atmosphere we breathe, or detect its fetid odor, this is no more a reason why we should set at defiance all experience, both personal and scientific, than it would be for a man who would swallow a poisonous drug, merely because he could perceive no difference, either in color or substance, between it and a cup of tea.

With these facts before us, establishing most clearly the fallibility of our senses, and their liability to lead us astray, it obviously becomes necessary for us to go to some other source for reliable knowledge in relation to the subject in hand. We find this knowledge furnished by science, in its interpretation of natural phenomena; so let us go to that with our minds divested of all prejudice, and with a firm determination to seek only after truth.

CHAPTER III.

WHAT VENTILATION IS.

VENTILATION is said to be of two kinds—*natural* and *mechanical*. With the latter mode we have nothing to do, as even if it were the better way, it can never, from its expensiveness, be made available for the "million," which is the great object I have in view. It would be no difficult matter, however, to show that mechanical ventilation can never be made as effectual as a natural or spontaneous ventilation. It obtained in an early period, before the natural laws which govern the motions of air were inquired into; and such is the effect of the prejudice and ignorance derived from those times, that it is still extant in England, and America also, to some extent. In this mode fans, blowers, pumps, etc., are set in motion by steam, or any other convenient power, and the air propelled or drawn in such direction and quantity as may be required.

The ventilation of a building is ordinarily accepted to be, the removal of the foul air from within, and its replenishment by pure air from without. This is true, so far as it goes; but several other points, of hardly less importance, must be taken into account also. We must not only effect a change of air; but in our climate, and particularly in the winter, the requisite temperature must be imparted to it, and it must be moved through the building in such a manner that currents of all sorts shall be avoided, so far as possible.

Our dwelling stands at the bottom of an ocean of air forty miles deep. We erect brick, stone, or wooden walls about us, and therefore we must find means to get so much of this ocean of air through the building as may be sufficient for our purpose. By opening windows, doors, or other apertures on opposite sides of our house, we can probably get a current of air through the building at certain points. Now, even if we could stand this in stormy or winter weather, and by night and day, still it would not be ventilation, because there would be only a partial removal or change of air. Every particle and atom of air must go out, or of course the process is not perfect. Currents of air, however, through a building, even if they would change the air — which they certainly can not do, be the process ever so ingeniously contrived—can not be tolerated in this cold climate, especially in winter; which is the very time when we need the most ventilation. The whole body of air, in each apartment, must move together, and every local current necessary to its motion must be so guarded and concealed that no inconvenience can be felt by the inmates.

There is no possibility of securing this motion of the air by any side or lateral movement; hence, as the only other movement is a vertical one, it follows that the entire mass of air must go either upward or downward; and such is actually the case.

In both winter and summer, the coldest or coolest air will, as we all know, lie at the bottom of any body of air—say a room-full—and will spread itself in level strata over the whole floor; and by the same law, the hottest or warmest air will, in the same way, lie next the ceiling of the room; the intermediate space, from the top to the bottom, being filled with strata, each occupying a position above or below, according to its temperature. Now, if this body of air could be *seen*, the lines of temperature would be observed to lie perfectly parallel, and perfectly level, from side to side of the apartment; so that two thermometers, one on each side of the room, at the same hight from the floor, would stand at exactly the same point. Now, as we can only move the *mass* upward or downward, and as we wish to move the whole mass, in order to avoid currents as much as possible, the question arises at once: What do we wish to secure in our ventilation, aside from the change of air? the cooling or warming of the room?

If we want to ventilate our room, to cool it, we must let the air out at or near the top, and supply its place with cool air, which, of course, will distribute itself over the floor of the apartment; and this has been the policy in nearly all our former modes of ventilation; cold air is introduced, which, taking up heat from the occupants of the room, and from the fire, immediately escapes, through an aperture provided for the purpose, at or near the ceiling. Thus, proceeding on the erroneous notion that cold air only could be pure, they have actually been freezing the people, when they wanted to warm them.

If, on the other hand, we wish to ventilate our house, to warm it, we must take the air out at or near the bottom, thus keeping up a continual exhaustion of the cooler air; and if we wish to set the body of air in the room in motion, upward or downward, we must of course bring in the necessary amount of outside air to do it. If we want to warm the room, the air we bring in must be warm; and if to cool it, it must be cool. It depends now entirely upon where you open the aperture to let the air out whether you can set this body of air so in motion or not. If you open the aperture at the top, and the air you bring in is warm—or if you open the aperture at the bottom, and the air you bring in is cold—in either case, the body of air will not budge; your warm air will go *through* the body, straight to, and out of, the top aperture, and the cold air will do the same, through the bottom aperture. The consequence of this state of things is easily seen—you will neither warm nor cool nor ventilate your room. But if you want to ventilate your room, to warm it, and open the bottom aperture, you will succeed in both; and if you wish to ventilate your room, to cool it, and open the top aperture, you will accomplish that, because in the first case the fresh air will be the warmest, and will not stop until it comes in contact with the ceiling, where, spreading out in level strata over the whole ceiling, it will keep its relative position to the whole body, until it reaches the bottom, and passes out of the aperture; and so of the cold air, if you open the top, and let the air out at that point. In both cases, every particle of air must be removed from the room; because, as air of one temperature can not, by any natural means, be made to move or stop out of its level, it follows that every *particle* of every *stratum* must in its turn leave the apartment.

Nothing short of this, therefore—the movement of the whole body of air in the apartment, and that *vertically*—can be called ventilation. It matters not, so far as the mere change of air is concerned, at what point of the room you let air *in*, it will immediately, and without mixing with the surrounding air, seek the zone or stratum nearest its own temperature; and hence, in order to save both time and temperature, if you wish to warm an apartment, the warmed air should be introduced at once at the top.

CHAPTER IV.

HEAT.

You can not contravene the laws of nature; but if you will work with, and assist her, she will cause a whole building, and every apartment, to inhale pure air, and exhale that which has been vitiated, just as naturally as the lungs of an animal do. Heat plays an important part in these operations, not only as a warming agent, but as a means for securing and continuing this healthy action; it is therefore proper that we should devote a little space to the consideration of its relations to ventilation.

Cold is merely a relative term; it has no existence as a positive agent or force, and is simply an absence of heat. Of heat we know nothing, except through its effects. What it really is, has puzzled many a wise philosopher to satisfactorily explain; therefore we shall not presume to theorize on the subject. As stated in the preface, we only have to do with practical matters, with just so much of science involved as will aid us in our explanations. We know that heat enters into the substance of all bodies, producing greater or less results according to its intensity. We also know that its tendency is to pass from one body to another, until they all become equally heated; and the colder a body is when brought in contact with a hot one, the faster will it take up heat. For instance, take three bars of iron, one at the ordinary temperature of the air, the second just bordering on a red-heat, and the third white-hot; place these together, the hottest in the middle, and the cold bar will receive far more heat from it than the one nearly red-hot. This proves that the colder a body by which you seek to extract heat from a hot one, the greater the quantity that will be obtained. Another illustration may assist in bringing out the idea more clearly. Take three pieces of sponge, let one be perfectly dry, another about half-saturated, and the third perfectly filled with water; bind these together, and that which has the least water, namely, the perfectly dry one, will receive more from the one fully saturated than the partly filled sponge will. We all have experienced the fact, that we lose heat from our bodies much more rapidly in a very cold day than in one not so cold; which is accounted for by the increased difference of temperature between our bodies and the air.

This passage of heat from one body to another is called radiation; and it not only occurs in the case of solid bodies, but fluids and gases also radiate heat, though to a much less extent than solids. It is radiant heat which warms us from stoves and fire-places.

Bodies are said to conduct heat, when they will take it up, and transmit it from particle to particle, the whole substance becoming heated. For instance, if one end of a copper or iron rod is held in the fire, the heat will gradually creep along through the metal, until the rod becomes heated the whole length. Solid bodies conduct heat much more readily than liquids, and liquids more readily than gases. Air, in fact, is the poorest conductor of heat known, being almost absolutely a non-conductor, unless it contains considerable moisture. Air, however, although it does not conduct heat—that is, transmit it from particle to particle—will carry it by its own motion. For example, if you bring cold air in direct contact with a hot cannon-ball, each particle of air receives more or less heat, and becoming lighter in proportion as it is heated, immediately rises, giving place to other particles, which, receiving their share of heat, also rise, thus creating an upward current of warm air from the hot body.

Now, this is just what we want to get at. We can not heat the air by radiation, for the heat-rays will pass directly through, without warming it a particle; and as we wish to warm the air, in order to warm the apartment through which it is to pass, the only way left us is to bring the air in direct contact with some hot substance, so that it may take up heat in the manner just mentioned, and then pass in through the building we wish to warm.

This brings us to the question of fuel. What arrangement shall we construct, by means of which we can heat the proper amount of air to the desired temperature, and have as little waste of heat as possible?

It has for many years been an impression on my mind, as I have no doubt it has been upon the minds of most inquiring people, that there was a great deficiency in the generation of heat, as well as in its application to the warming of our buildings. In short, a great waste of our fuel.

We should consider that it is the air in the room that is cold—nothing else. The question, then, is: Which can I do the cheapest; warm the air where it is, or get rid of it, and replace it by other air that shall have been warmed?

All our aim, all our experiments, and all our practice, have hitherto been predicated upon the supposition that the body of air within the room was to be heated. Down to the year 1843, when I first began my experiments, our fire-places, our stoves, and our hot-air machinery, were all directed to this end, namely, the heating of the air already in the room. If we have but one apartment, and it is the mere heating of this that is required, I believe that the most economical way is by the common stove; but when we have several apartments, the cheapest way of warming them is by means of other air. Indeed there is no other way of effecting this object — the warming of several rooms from one source of heat, (unless that source be placed in each apartment,) than by substituting other and warmed air, for that already within the room.

I once thought that several apartments might be warmed from the one in which the stove stood, and instituted many experiments predicated upon this supposition; but for all practical purposes they turned out failures. By making apertures at the top and bottom of the division-wall, between the stove-room and the cold room adjoining, a change of temperature would take place; but, notwithstanding the stove was made red hot, and the room in which it stood much too warm for a person to live in, yet the adjoining room, in a zero day, could not be brought up to over forty-five degrees at the centre.

This I found was entirely owing to the want of a sufficiently rapid circulation. The difference of temperature between the top and bottom of an apartment inducing vertical movements more or less rapid according to its height, together with the direct radiation from the hot metal in the body of the immediately surrounding air, will warm a room, but it is doubtful whether the mere circulation derived from a difference of temperature merely, would be sufficient to warm *it*, much less, an adjoining room. Indeed, my experiments have convinced me, that it is impossible to warm several apartments by one common stove, manage it how you will. If, therefore, we want to warm several apartments from one source of heat, we must induce a circulation, much more rapid than can be had by means of the mere difference of temperature in the room. In a word, the air must be drawn *entirely out of the house*, in order that a more rapid circulation may be had.

The removal of the air, being the very process required for ventilation, it follows that it is cheaper to warm a building containing more than one apartment, by the ventilating process than by any other.

Air is the only means that I know of, the only medium by which heat can be conveyed from one apartment or place, to other and distant rooms; but this it will not do unless you put it in motion, and this motion you can not produce without exhaustion by means of chimneys or other air-shafts. The air, after it has received its load of heat from your warming machine, must, by mechanical means, be directed to the apartment in which it is required, not by means of any pipes or conductors, but only by the force of the vacuum created within such apartment, by the exit of so much of its air up the shaft, as is to be drawn in from the air-warmer. An aperture must of course be left by which the warm air can enter the room, and as before stated, this should be near the ceiling. Then, like a good errand-boy, the air is valuable in proportion to the celerity of its movement, as the cold walls by which it is surrounded are constantly robbing it of its heat.

Time in this operation is every thing, because the moment the warmed air leaves the top of the air-warmer, it begins to lose its temperature, and increases in this tendency, according to the time it remains in the apartment which it is to warm.

As time then is so important, it behoves us to make such provision by means of our mechanical arrangements, as will empty our apartment as quickly as possible.

So great is the difference of locality, and size of different apartments in different houses, that it is utterly impossible to lay down any *rule* or *formula* by which to be governed in the generation of heat, to make them comfortably warm in cold weather. People are generally impressed with the idea, that it is the ground area, or lateral measurement of a room by which they are to be guided in arranging for its warming, but this is not so; it is the *hight* by which they should be chiefly governed. When I tell you that it takes about double the quantity of fuel to warm a room twelve feet high, that it takes to bring one of the same lateral measurement and ten feet high to the same temperature, you may be surprised, yet this is a fact which I have proved by many experiments. Here lies the capital mistake in our dwellings in Canada. However, I shall treat this point more particularly in another place.

To sum it up, then, we want our houses warmed and ventilated. For its ventilation, we must supply a brisk and easy flowing body of air, and the oftener this is changed, of course the less heat it is necessary to impart to it, as it comes in, that is, until you get down to a certain point. The air should pass out of the room at about 60 degrees, and this will make it necessary to take it from the air-warmer at 90 or 100 degrees in an ordinary winter's day, and with an ordinary sized house, say forty feet square. In a very cold day, when the air is liable to lose its heat much faster, from the increased conducting power of the walls of the building, it should be brought in at a somewhat higher temperature.

I think no one will venture to deny, but that the ventilation is necessary, and if this be the case, there certainly is great advantage in being able to bring the air in at that comparatively low temperature, as thereby all danger of injuring it by over-heating, as is too often the case with hot-air furnaces, is evident; and there is also an obvious saving of fuel, for admitting we must have a certain quantity of air-way for the purpose of ventilation, it certainly does not take so much heat to raise this from freezing to 90 or 100 degrees, as it does to raise it from freezing to 200 degrees, or as is often the case with furnaces, to 500 degrees, especially where ventilating arrangements for the outflow of hot air are provided near the ceiling. The reader is referred to Chapter V. for a description of the air-warmer.

CHAPTER V.

THE AIR-WARMER.

Strange to say that the sun does not warm us as well in winter, when it is quite near us, as it does in summer, when it is many millions of miles further off. The reason of this is obvious enough, after a little reflection, and an appeal to our every day's experience. To dry or warm a board, we do not set it edgewise toward the fire, nor partly edgewise; but we hold the flat surface to the heat. The reason why it dries or heats sooner in this position is, that the rays of heat strike upon the surface direct, and can not glance off, so with the rays of the sun in summer; in our latitude, they strike upon the earth more directly than in winter. In the tropics, where the sun is nearly vertical at noon, it is so hot that a human being can not long exist when subject to its rays; whilst in a very high latitude, far north, its influence is scarcely felt, not because of its distance, but because its rays strike the earth obliquely.

Rays of heat, whether proceeding from the sun or from a common fire, act upon the same principle. It stands to reason, therefore, that if you wish to absorb the greatest possible amount of heat, in the shortest period of time, by a metal-plate, you will place this plate in such a position that the heat shall strike it *directly*, and upon the broadest surface; not edgewise or in a slanting direction.

It is obvious, then, that the generally received notion, amongst hot-air-furnace and stove manufacturers, to increase the surface, by means of pipes, corrugated and otherwise, is vicious, or rather, is unwarranted upon any philosophical or useful principles.

A cast-iron metal plate is very porous, and the heat runs through it as water through a sponge or bag; indeed, it has been doubted whether, by its equalizing and concentrating power, heat will not pass through it into a cold body of air outside, as readily and rapidly as if the plate did not intervene.

The proper *shape*, therefore, of the metal of an air-warmer, is that which presents the greatest surface to the centre of your fire; and that shape being globular, it follows, that this is the form most efficient and economical for warming our houses by means of air. Where wood-fuel is used, this shape can not so conveniently be strictly adhered to; but the two sides of the air-warmer, after a perpendicular rise, high enough for the fire-chamber, may be gradually made to approach each other toward the top, which will preserve the globular form, so far as circumstances in such case will permit.

Not only is it necessary for the metal, forming the inside of the warmer, to present toward the heat a plain surface, but it is just as important that the outside, over which the air is to pass, should also be plain, thus enabling it to pass the rays of heat, in the most direct way possible, without any intervening indentations, corrugations, or other surfaces which will tend to throw out of their course the rays of heat.

We feel cold, not because the cold air strikes into our bodies, but because it takes or extracts the heat from the body; and it is because our woolen fabrics, being the best non-conductors of heat, prevent the heat from leaving the body, that we call them the warmest, and not because they keep the cold out.

Heat, as we have seen in the case of the hot and cold bars of iron, passes from one body to another, more or less rapidly, according to their difference in temperature, until their temperatures become equal. You may have been unduly warmed by walking, and meeting a friend, on a cold, windy day, to whom you have something to say, you stop, and in a few minutes experience a disagreeable sensation of coldness. You retire with your friend under the lee of a wall, and you feel comfortable. Now, a thermometer, held out in the wind, and then under the wall, will in each case indicate the same temperature; hence the coldness was produced by the heat being conducted away by the wind; the particles of cold air succeeded each other so rapidly, as to prevent any of them from taking much heat; but the difference of temperature between the air and body was continually kept at that same low point by the constant succession of cold particles, and these extracting heat with a power proportioned to the rapidity of their motion and the lowness of their temperature. If you hold your arm or leg out from the shelter of the wall in the wind, it will rapidly lose its heat, whilst that part of the body protected by the wall still feels comfortable, as the air is at rest around it, thus preventing the heat from escaping at so rapid a rate as in the other case.

Now, if we apply this reasoning to the extraction of heat from the hot metal-plates of an air-warmer, we shall be convinced, that the more rapidly we can bring the air over the heated metal, the more heat we shall extract from our fire; and this is exactly what we want to discover, in order to shape our heating-machine.

It only requires a cursory view of our hot-air furnaces, with bulb upon bulb, and pipe upon pipe, laid horizontally perhaps, to see that neither of these objects can be obtained. The only rapid current of air within their hot-air chambers is around the outside of the furnace, touching merely the edge of the bulbs, and so passes off. To be sure there may be, and no doubt is, a little movement of the air *between* the bulbs and horizontal-lying pipes; but it is already so hot that it is of little use in the capacity of an absorber of heat.

Whatever air may come out from between these metal bulbs, and mix with the current outside of them, is completely destroyed for respiration; but mixed with the cooler current, is again brought down in temperature, and from the time it enters the distributing tin pipes, until it reaches the apartment it is destined to warm, is constantly and rapidly losing heat, perhaps by hundreds of degrees, until it *feels* healthy. The inmates of a house thus heated, do not see that air, after having been once rendered unfit for breathing by heat, can not be restored by mere cooling. It frequently happens that a register, or even all the registers, may be closed, and perhaps for hours, in consequence of the rooms getting too hot. A strong coal-fire is constantly doing its work of destruction to the life-giving principle of the air, within the hot-air chamber, and of which, when the registers are again opened, the inmates receive the full benefit. Moreover, as is well known, there is an elimination of sulphur from brick and stone walls, when heated, as well as an absorption or destruction of the natural moisture of the air, thus putting it in a condition, when taken into the lungs, to produce head-aches, dryness of the throat, and inflamed eyes. Is it not a matter of wonder that this mode of *heating* our dwellings has been so long endured?

It is no part of my plan, however, to find fault with other modes of ventilating and warming houses; I only wish to enforce my own.

I admit of no hot or warm-air chamber except the building itself. The air-warmer, whether placed in the basement, or in the hall, or any other convenient place, should never be so covered as to admit of being closed, except to regulate the quantity of cold air, when there is a fire in it. If set in the basement, all the warm air they manufacture should at once be let up into the building, through a register, which will freely pass the whole quantity that comes through the air-warmer, thus filling the whole house, each apartment drawing on the whole body of air, according to its requirements. This operation will hereafter be particularly described and exemplified.

CHAPTER VI.

AIR AND WATER.

It is supposed that the atmosphere surrounding the earth extends upwards to a height of about forty miles. It is estimated, that every square inch of surface, on a level with the earth, sustains a pressure from this body of air of fifteen pounds. Thus the body of a man, the surface of which amounts to two thousand square inches, will sustain a pressure from the surrounding air, equal to the enormous amount of thirty thousand pounds; and if it were not the nature of fluids to transmit pressure equally in every direction, of course every thing which could not sustain such a pressure, would be crushed; but such being the case, the internal pressure being exactly equivalent to the external, we move about freely, without becoming aware of the enormous weight that our bodies continually sustain. The power and weight of the atmospheric pressure, at the surface of the earth, will sustain a column of water thirty-four feet in height; that is, supposing such a column, no matter of what diameter it may be, be inclosed in a straight tube, open at both ends, and the pressure of the air taken from the top, by an air-pump or any other means, it would be sustained there by the pressure upon the lower-end surface.

Whenever heat is applied to water, air, or any other fluid, or gaseous substance, the immediate effect is motion; and the operation of what we term a draft or upward motion of the air, is exactly similar to and governed by the same laws as boiling or heating water. The heat at the bottom of the boiler warms, or in other words rarefies and renders lighter, that which lies next to it, and it forthwith rises to the top of the body. But it must be observed, that this rising is caused no more by the lightening of that portion of the water, than it is by the comparative increase of pressure of the surrounding body of water downward. Indeed, these two motions are and always must be simultaneous, for the obvious reason, that there can be no such thing as any natural vacuum. Precisely the same effect would follow, if, instead of rendering lighter any portion of the water, we should increase the weight of that which surrounds it; namely, a motion upward. Thus: take some solid body, but a little heavier than water, bulk for bulk; this will, of course, sink and lie at the bottom of a vessel filled with water; now take another body or substance, heavier than water, and capable of being held in solution, or, in other words, dissolved or liquified by it, say common salt; when this is added, the body at the bottom of the vessel will rise, from the increased pressure of the surrounding body of water holding the salt in solution.

It is to a similar operation in the atmosphere, which takes place under similar circumstances, that we are mainly indebted for the draft in our chimneys. It is this principle in the atmosphere, which causes a pump to work; and not, as is generally supposed, a suction. The drawing up of the upper valve, merely takes the weight of the atmosphere from the top surface of the water within the pump; and leaving it upon the surrounding body of water, contained in the well, the water within the pump is forced upwards, upon the same principle as the thirty-four foot column of water, of which I have been speaking, is sustained. To convince you of the correctness of this statement, (and it is important in the working out of the ventilating principle, in regard to the weight of the atmosphere, however strange it may seem to some,) try to pump water out of a well, where the water lies more than thirty-four feet below the surface, with a force of less than seven hundred and seventy-two pounds in power, (this being about the atmospheric pressure upon a four-inch diameter round bore,) and you will fail. You will further find, that your fixed valves or boxes must be placed within the tube of the pump, *within* thirty-four feet of each other, at from whatever depth you wish to raise the water—a clear proof that the atmosphere (of forty miles) is as heavy as thirty-four feet of water, the area of the columns being equal.

And any thinking, practical man, who is familiar with the power of even a ten-foot waterfall, in its effect upon the water-wheel of a mill, will at once comprehend the power in a fall of thirty-four feet.

So also, I may mention, is the operation of our breathing. By an enlargement of the cavity of the chest, produced by the operation of certain muscles, the air, by its superincumbent weight alone, is forced into the lungs; and this, what we call inspiration of air, is *not* the result of any power we can exercise over it, as is generally supposed. It is estimated that one half of the whole weight of the atmosphere is contained within the first three (out of the forty) miles, and hence, as a further proof of the weight and power of the air, a great difficulty of breathing is experienced on the tops of high mountains—one half of the pressure on the lungs being taken off at that height.

Water and air are both fluids, and in their operations are governed by the same laws; they both boil by the application of heat; they both have weight; and will both run or take a *downward* course, if not prevented by some obstruction. They will both seek and obtain the lowest place, in regard to any body with which they come in contact, which is lighter or of less weight, bulk for bulk. The subject of the ventilating system will therefore perhaps, be better comprehended, if we keep in view the natural operation and course of water, with which, as it is visible to the eye, we are all familiar.

We will suppose, then, two rooms, one placed over the other, the lower room filled with cold, and the upper with warm water. Open an aperture *between* them, and neither of them will thereby change, either their places or temperature, because cold water is heavier than warm water. Just so with warm and cold air; the latter will always be found at the bottom of a room, because it is the heaviest. Now open a gate in the bottom of the cold-water room, and let its contents run out, and the warm water from above will occupy its place. Just so again with air; the cold must first run out, drawing the warm air after it. The water takes a downward course, as it runs out of the lower room, and the air an upward direction; why? simply because the water being about eight hundred and fifty times heavier than the air—the medium it has to contend against—takes its lowest and natural place; but convert this water, as it runs out, into a lighter body than the air, say into steam, and it will immediately go *upward*—the air, now being heavier than the water, will keep the lowest place.

Now, with these facts in relation to air and water, let us suppose that we have erected the dwelling-house, represented by Plates 7, 8, 9, 10, and 11, and that it is finished to the turn of the key; and that it stands at the bottom of Lake Ontario. Let us suppose, further, that the foul-air shaft is reversed; instead of going up it goes *down*, reaching through a supposed crust, forming the bottom of the lake, all below being infinite space. Now what we want is, to produce a perfect water-ventilation, and it will resemble exactly what we are at, in trying to get an air-ventilation.

We must, to begin with, suppose the whole building to be an air-tight tube, of which the fresh-air duct B, is one end, and the foul-air shaft, (now turned downward,) is the other end; for it is none the less a "tube," because it may have been enlarged in the middle, and divided into apartments, such as dining-rooms, drawing-rooms, etc.; it is still airtight, and not a drop of the external water can come in, except when let in by the fresh-air duct B, (which must be the case with every building erected for ventilation, as nearly as human ingenuity can do this.)

The whole being now air or water-tight, let us suppose that the valve, in the fresh-air duct B, as also the fresh-air regulator, Fig. 3, Plate 52, under the air-warmer, as well as all the fan-registers, F, Fig. 3, Plate 3, to lead the air into every apartment, and the four sliding-valves, F, Plate 10, together with the base, Fig. 1, Plate 3, are all open. It is now obvious that the valve in the fresh-air duct B being opened, a constant and rapid change of water is going on throughout the whole building, and every apartment within it, to the full extent of the capacity of the fresh-air duct B, to let it in. Let us follow up this operation. The water enters at B, proceeds up through the air-warmer—is warmed—fills the hall, (note here, that it can not get under the hall-floor until after it has first been through some one or more of the apartments, and flows into the space under the hall-floor, through the apertures, covered by the valves F, and which space I have elsewhere called "the horizontal part" of the foul-air shaft,) flows into the fan-register, runs down through the open base, through the apertures of the valves F, under the several apartments, and into the space E, under the hall-floor, (which space I call the horizontal part of the foul-air shaft,) thence in the direction indicated by the arrows into and down the foul-air shaft into space.

We must now imagine the whole building filled with water, and, as above stated, a rapid change of water going on, as long as we leave the fresh-air duct and all the valves and fan-registers open.

We will now take another view of this operation, and in doing so we must not forget what I have elsewhere stated, that you can not, by any natural means, take an atom of air *out* of an air-tight apartment, unless you let exactly that same quantity into it, nor can an atom come in, unless that same quantity be taken out. If, therefore, we now close the fresh-air duct B, it will just as effectually stop the circulation of the water in every part of the building, as would the closing of the foul-air shaft going down into space; and yet the building will remain perfectly full. So, if you cut off the connection at any point, between the points of ingress and egress of water, even supposing the fresh-air duct and foul-air shaft are both left open; for example, if the four valves F, under the four principal rooms, be closed, the connection between the fresh-air duct B and the foul-air shaft is cut off, and not a particle of water can either come into or go out of the building any more than in the other case.

If, now, we wish to set the ventilation a-going, we pull up the valve F, for instance, under the drawing-room; immediately, just the quantity of water that the aperture covered by that valve will admit, will at the same instant pass up through the air-warmer, through the fan-register, for the drawing-room, through the open base, and out of the foul-air shaft. So also will the ventilation in the other apartments be resumed by the same means when required.

The only difference you will find, between this water-ventilation, which I have endeavored to show you, and the actual air-ventilation, is, that in this case the building and its several apartments are filled from the bottom upward; whilst in the air-ventilation, the building and its several rooms will, when the air is warmed in winter, be filled from the top downward. The foul-air shaft, in this case, carries the water downward, whilst it will draw the *air* upward.

Of course, when there is a basement or cellar under your building, the firring of the joists, the foul-air-gathering duct, and the lathing and plastering of the ceiling of the basement must be attended to.

CHAPTER VII.

CELLARS.

There is great diversity of opinion, in regard to having cellars under dwelling-houses. There is, certainly, where room is an object, economy and convenience in it; but I believe there is disease and death in it also. It is true, that a building properly ventilated will in general prevent any miasm from coming up into the rooms above; but the vast quantity of foul air generated in these receptacles of meat, fish, vegetables, milk, butter, cheese, and the many other edibles and commodities which find their way into these apartments—all, or most of them, however fresh, in a state of decomposition and decay, will pervade any unventilated dwelling, and in spite of all ventilation, now and then find its way up the stairs through the joints of the floor-boards, and even through the timber itself.

The enormous quantity of carbonic acid gas, and other foul air, constantly engendering and accumulating in some cellars, not only comes up the stairs, and through the cracks and crevices, but through the timber itself, saturating it, as well as the brick and stone walls with which it comes in contact. The woolen carpets, curtains, wall-paper, and other furniture, in an unventilated house, become so thoroughly soaked with this mephitic air, that no amount of scrubbing, washing, or cleaning can eradicate it; the whole building, in fact, becomes one mass of putridity. The sun never reaching it, the cellar, if not actually what we should call wet, will yet retain that sort of humidity which adds virulence to the putrefaction constantly going on.

We have, indeed, sufficient premonition of the unhealthiness of cellars, for we never enter one but we are sensible of an unusual and nauseous smell. This faculty of smell is not given us for nothing; if we disregard it, we can no more expect to escape the consequences than we can a willful disregard of our other senses. This applies equally to the same sensation in going into a room in an unventilated house. The reason of our greater disregard of our smelling attribute is plain enough; we are so accustomed to meeting with different odors in different houses and rooms, and having never found any cure for it, we quietly submit to what we deem an inevitable necessity—and so the matter drops. The fact is, that whenever and wherever our olfactories are sensibly and disagreeably affected, there is something wrong, and it immediately becomes our bounden duty to remove the cause of it if possible, more especially, when we know it proceeds from the decay and putrefaction of animal matter.

It is utterly impossible to keep cellars perfectly dry, whatever their locality. In summer they are cooler than the surrounding air; and whenever it finds its way into them from the outside, an immediate deposition of moisture takes place, from the change in temperature that it undergoes. The sun-light can never be admitted to exert its purifying influence, and so the moisture goes on accumulating. In winter the cellar is stored full of vegetables of various kinds, which, by their gradual decay, impart moisture to the air within, as well as carbonic acid and other noxious substances; and as the cellar is usually tightly closed during the whole winter, the air becomes damp and offensive, without any chance for getting rid of it until the cellar is opened in the spring. These things will occur in cellars in the driest localities; and in damp, clayey soils, there is not one in fifty but what has more or less water standing in it, unless provided with an efficient drain, and even then it can never be made dry enough to live over with impunity.

My observations have convinced me, that although children born in dwellings, with such cellars under them, may attain to maturity, yet they seldom become robust and healthy men and women; having, however, if I may so express it, become acclimated, they endure much longer than those who may perchance succeed them in the occupancy of the same building. In this case, typhoid and other fevers are almost invariably the consequence; but if the cellar is a naturally *wet* one, which is the case with a majority of them, consumption is almost the inevitable consequence to the new occupants.

I have not a doubt but that a vast amount of disease arises from families moving into and occupying dwellings which have *wet* cellars under them, especially where the houses they left had comparatively dry cellars; and I may here add, that many years' observation has convinced me that it is dangerous for any family to move into any unventilated dwelling, in which a sickly family has lived—no matter however remotely.

In relation to building houses without cellars, Dr. Buchanan says: "While I would condemn cellars and basements entirely, the common plan of building, in their absence, must be condemned also. The house being built above the surface of the earth, a space is left between the lower floor and the ground, which is even closer, and darker than a cellar, and which becomes, on a smaller scale, the source of noxious emanations. Under-floor space should be abolished, as well as cellars and basements. The plan that I have adopted, with the most satisfactory success, to avoid all these evils, is the following: Let the house be built entirely above the ground; let the lower floor be built upon the surface of the earth, at least as high as the surrounding soil. If filled up with some other material, a few inches above the surrounding earth, it would be better. A proper foundation being prepared, make your first floor by a pavement of brick, laid in hydraulic cement, upon the surface of the ground. Let the same be extended into your walls, so as to cut off the walls of your house, with water-proof cement, from all communication with the moisture of the surrounding earth. Upon this foundation build, according to your fancy. Your lower floor will be perfectly dry, impenetrable to moisture and to vermin; and not a single animal can get a lodgment in your lower story. By adopting this plan, your house will be dry and cleanly; the atmosphere of your ground-floor will be fresh and pure; you will be entirely relieved from that steady drain upon life, which is produced by basements and cellars; and, if you appropriate the ground-floor to purposes of store-rooms, kitchen, etc., you will find that the dry apartments, thus constructed, are infinitely superior to the old basements and cellars. And if you place your sitting and sleeping-rooms on the second and third floors, you will be as thoroughly exempt from local miasms as architecture can make you."

While I am strongly in favor of abolishing basements and cellars altogether, under dwelling-houses, still, in order to have a thorough system of ventilation and warming, a current of warm air should be passed under the floor. If a house was to be built, therefore, with the intention of carrying out the warming and ventilating policy, it would be necessary to leave a space between the brick pavement and the underside of the floor timbers, say of two or three inches, in order to get a free circulation of air; and, as long as this air was kept in motion, all the advantages, mentioned by Dr. Buchanan, would be still retained; besides, the timber would be effectually preserved, and nothing hazarded by the open space.

CHAPTER VIII.

WOOLEN CARPETS.

One point which above all others demands our attention is, that abomination of civilized society, woolen carpets. Their use is very injurious in a thoroughly ventilated house; but ten times worse in ordinary houses, where no provision for ventilation has been made. The fact, that, in carpeted rooms, we are living enveloped in an atmosphere of dust, is sufficiently proved even by the ordinary sight. The rays of a rising or setting sun, accidentally entering a window, frequently exhibit this to the naked eye, and to such an extent, frequently, that the beholder moves involuntarily away, as he supposes, for the moment, from its influence. The whole room is filled with it, and it is in constant motion; and so long as woolen carpets are in use, and any of the present modes of heating houses be persisted in, so long will the inmates of such dwellings be subject to this health-destroying respiration.

The moment you place fire or heat in the centre of a cold room, having no open flue in it, that moment, every particle of air within that room is put in motion. This motion is upward, from the centre of heat, and rotary, similar to the water in a boiler or cauldron placed over a fire, rising from the center to the top, thence outward, and down the sides of the boiler, until it again reaches the spot it started from, and so on. The hotter your stove gets, the more rapid will be the ebullition. Every step taken upon the carpet, especially when near the centre of such a room, a quantity of impalpable dust is sent to the ceiling, until the whole room fairly becomes hazy. As proof of all this, you have only to examine the tops of your book-cases, window cornices, or shelves of any kind, which will be found covered with dust; and, in the best kept room, with a woolen carpet, you may write your name every five minutes in the day upon the furniture, especially if it be placed near the walls of the room.

In support of what I have here advanced, I will offer the following extracts. In the Proceedings of the British Association for the Advancement of Science, we find the following communication, from a medical student, on a disease of the lungs caused by the deposition of particles of dust. It was read by Dr. McIntosh, and would contribute, he observed, toward the elucidation of that class of diseases affecting artisans, which had, in a more systematic form, been treated by Mr. Thackray: "In the neighborhood of Edinburgh, were many stone quarries, the workers in which not unfrequently died of consumption. A mason, a worker in the Craigleith quarry, was ill, he was bled and treated for a common cold, recovered and returned to his work. A short time afterward he was again taken ill, and two years after the first attack he died."

The account goes on to give the details of the appearances of the lungs on a post-mortem examination, and adds: "He directed particular attention to this analysis, for Dr. William Gregory had published an account of the Craigleith quarry-stone, and the analysis of this stone gave the same ingredients as those found in the lungs of this workman. Dr. Gregory found in the stone carbonate of lime, silica, and alumina. The deduction must necessarily be," he adds, "that this must be an absolute deposition of the Craigleith quarry-stone from small particles taken into the lungs during respiration, producing consumption and death."

Dr. M'Cormac says: "The habitual respiration of foul, unrenewed air I look upon as the only real source of tubercle, including under this designation both phthisis and scrofula. Unless foul air be respired, there can be no consumption, no scrofula. If an individual live constantly, day and night, in the open air or in air of equal purity with that subsisting in the exterior atmosphere, he can not incur consumption. There are no consumptive Gipsies or Bedouins, so long at least as they preserve their aboriginal or out-of-door usages, or are not subjected to confinement or ill-treatment. As for hereditary consumption, making due allowance for the few individuals born tuberculous, and for the greater proneness under like circumstances of those sprung from diseased progenitors, to disease, there is no such malady."

In one of his lectures on the use of the lungs, Dr. Fitch says: "Inhaling or drawing in large quantities of dust will cause a deposition upon the lungs, and thus by mechanical irritation lead to consumption. This is seen in stone-cutters, millers, dry-grinding of metals, pickers and sorters of rags for paper-making, and many others. I once knew a case of a stone and marble-cutter who died suddenly. His chest was opened, and it was found that a large proportion of both lungs was so impregnated with stone-dust as to have caused his death. This occurred in Cincinnati, Ohio."

Dr. Dixon says, speaking of carpets: "It is now well known by those who investigate by dissection, after death, the diseases of the body, that the microscope will frequently detect small particles of wool, and minute concretions of particles of dust, inhaled from the atmosphere, where they are constantly flying about. The lungs, in a state of health, will ordinarily throw off all such matter by sneezing or coughing; still, as it has repeatedly been found in the lungs of persons who have died from pulmonary disease, confining them to the house for weeks anterior to the period of death, the inference is irresistible that they have been inhaled from the apartment in which the sick person has lain. It may be said that this would exclude altogether the use of carpets from sick-rooms, or indeed from the whole house. Strictly and logically speaking, it would; but owing to custom, and the natural power of wood and stone, by means of their superior density, to conduct off the animal heat, and that with distressing results in debilitated persons, and as our women can not be educated in this generation up to cork-soles in the house, carpets will doubtless continue to prevail till a more elegant and artistic taste shall banish and replace them with hard or polished wood. Very well, then, as what can't be cured must be endured, let us investigate the properties of those gorgeous carpets with the fleecy and velvety textures, which are so indispensable to the happiness and gentility of an American housekeeper. Every one knows that they can not be thoroughly swept, because the dust is forced downward at each successive effort, and protected from the action of the fibres of the broom — this is evident enough; moreover, as they are tacked down, and are exceedingly heavy, they are rarely shaken, whilst the windows are kept shut as much as possible, to avoid the dust from the streets. Whoever enters a parlor thus carpeted, will perceive by the peculiar character of the close and oppressive smell, that it proceeds from the carpet. Now, in one word, the more there is of it the more this smell will prevail; if it be unhealthful, the longer the parlor is occupied the greater will be its evil influence on the lungs."

If our ordinary sight were equal to it, we should regard carpeted rooms with perfect horror. A woolen carpet will last about a dozen years, and within that time our lungs have taken in almost three fourths of its original weight. What a pity that so beautiful an article of furniture should come to so ignominious an end!! Woolen carpets, brooms, and dusting-brushes, should be banished from every family. They produce and entail upon every succeeding generation greater mortality than war or pestilence ever did. No family can be a healthy one where they are tolerated. Oil-cloth coverings for our floors, if they must be covered, are much better in every respect. They may at first cost a little more, but they last longer, and the saving in doctor's bills will much more than balance that. But when we reflect upon the amount of misery they will prevent, for generations unborn, who that values health can hesitate to choose between the two? A mop and pail of water every morning will do in five minutes what, with a woolen carpet, will take Betty with broom and dusting-brush an hour, to say nothing about wear and tear and the discomfort of the family.

That carpets are fashionable I admit; but that is their sole recommendation. I have not the least doubt in my own mind, that to the health of persons using them they are the most destructive things possible, and that the sins of those who persist in their use will be visited upon their children to the third and fourth generations. Our ancestors were vulgar in their notions, in their language, dress and manner of living, according to our ideas in the middle of the nineteenth century; but where is the robustness, vigor, health and energy of character which distinguished those of the seventeenth century? This period of early dinners, wainscoted houses and polished floors!

Now I insist upon it, that a polished floor, or a floor covered with a well-kept oil-cloth — albeit the former may be somewhat more expensive — so far from being vulgar, would, in my humble opinion, be the very reverse. If a general or common use be the test of vulgarity, then I submit that a carpet comes preëminently within the category; for scarcely a house can be found which can not boast of its carpet, and ought, according to such reasoning, to be repudiated on that account alone.

CHAPTER IX.

COLD FEET.

In the year 1843, I discovered that air could not only be drawn from the bottom of a room, but that the draft thence was much stronger than from the top of the apartment. It appeared evident, therefore, that in any cold climate where artificial heat was required, inasmuch also as the air lying at the bottom of a room was always colder than at the top, this was the place at which it should be extracted.

I discovered, however, after a series of experiments, that, notwithstanding the air might with greater economical advantage, be taken from the bottom rather than the top of an apartment, yet in a climate like ours this process produced an evil not less important than that involved in a waste of fuel in extracting the air from the top of the room, namely, a strong local current over, and close to the floor from all parts of the apartment, and from the fact that the centre of the draft was *above* the floor, the quantity was much increased from the cold air always lying between the joists of the main or lower floor of the building, from under the skirting, and even from adjoining rooms, causing cold feet, which were thus immersed as it were in a cold bath. This sheet of cold air, moreover, converging towards the exit-aperture, rendered it impossible, without suffering great inconvenience in winter, to sit at any point in its vicinity. If to this be added the physiological fact that the extremities of a person, such as the hands, face, ears, and feet, are the more sensibly affected by cold than other parts of the body, in consequence, I suppose, of their greater distance from the source of heat, we can very readily see that the feet, instead of being subject to the influence of the coldest air of a room, should in fact be surrounded by the warmest; and, moreover, to *feel* warm, this air surrounding the feet should be inert, stationary. It is well known, by every experimenter on heat and cold, that a strong current of air warmed up to $70°$, $80°$, and even $90°$ will produce a sensation of coldness to every part of the body, whilst the same air in an inert state will feel insufferably warm. The blood is of a temperature of about $98°$ Fah., and any substance of a *lower* temperature than $98°$ will *feel* to the skin cold, and will feel colder still in proportion to the number of particles of the substance or material by which we are surrounded, brought in contact with the skin. The reason why we feel warm in a room up to 70°, whilst the temperature of the blood is 28° warmer, is owing to the fact that the air in a room being in an inert or quiescent state, and our clothing being generally of a non-conducting fabric, especially if of wool, the heat is obstructed in its passage from the body, and hence the body, or blood, which is the same thing, *retains* its 98° or nearly so. But put the air of this room at 70° in motion, either around the feet or any other part of the body, and thus increase the number of particles of air in contact with the body, and you immediately feel the heat leaving the whole frame. I say the *whole frame*, because, if taken from any part or member of the body, it is felt in *every* part. The clothing we wear to be sure prevents a considerable portion of heat from leaving the body, but yet the effect is instantaneously felt through the thickest woollen clothing, and the most substantial shoes or boots. Moreover, the feet being the most exposed to wet and cold, have always from mere instinct been more substantially guarded than any other part, and hence, these members, by means of wool and leather, have become more tender and sensitive to cold than any others. We have all observed how instantaneous is the effect upon the whole system, of putting our feet in hot or cold water. In the first case, if we are in ever so cold a room, we will feel a glow of warmth; and in the other, if we are in a hot room, we feel a cold chill like an electric shock to the very crown of the head.

The truth is, when our feet are warm, we feel warm all over, and on the other hand, when our feet are cold, we are cold all over, at whatever temperature the room may be.

If I have stated this case truly, it will go a long way in establishing and confirming what has long been my conviction, namely, that we have all this time been wasting our fuel and injuring our health by an endeavor to warm our *feet over our heads!* Our heads and upper part of our bodies have been obliged to endure 70° of heat merely because the cold was required to be kept from our feet.

Do not for a moment imagine, that I am an advocate for the application of any *active* heat to the feet. This I am aware has been condemned by the medical profession as pernicious to our health. All I want is, *to keep the cold away from the bottoms*, which is the sensitive part. If this can be accomplished, (and before I have done I shall prove to you that it can,) we will live more comfortably as to warmth, (to leave health out of the account,) with a general average of warmth in an apartment at the height of the head of 60° Fah., than we now do at 70°.

If such then be the importance of having the cold kept from our feet, it is evident, that a *current* of air, even warm air, must not be suffered upon the top of the floor; nor must the cold be allowed to penetrate upward *through* the floor.

I need not stop to use any argument at this day, to show that in this climate in winter, and in view of the scarcity and dearness of fuel, we can not allow the air of a room to be taken out at the *top*, and having already shown that it will not do to take it out at any *one* aperture at the bottom, some other means must be adopted by which it can be got rid of without the extravagance of the one mode, or the inconvenience of the other.

That the air *must* be taken *from the bottom* of the room somehow, admits of no doubt in my mind, and after much reflection and many experiments, I found that the local currents which were induced by the exit-aperture being above the floor, (by drawing the cold air upward,) might, by merely making the aperture *under* the floor, be transferred from the top of the floor, to the under side, where the currents could do no harm. In order, therefore, to allow all the cold which always occupies the spaces between the joists, and by which so much cold is communicated *through the floor-boards* to our feet, to find access to this aperture, I laid two-inch firring on the top of the joists, and laid my floor upon them, thus providing a space of two inches for the air to circulate *across* the joists, and consequently opening up a complete and free communication to the exit-aperture above mentioned. So far so good; this certainly would prevent any of the cold air from below or amongst the joists from coming *above* the floor at any rate. But now there was another point to be gained, and without which no warming or ventilation of the room above could be effected. This was the want of an exit-aperture for the extraction of the air of the *room* from above the floor. I therefore diminished my floor two inches all around the room, (except of course where the hearth or doors came,) and thus not only did the air of the room find an open and free communication with this same exit-aperture; but being warm air, would naturally impart its remaining warmth (which now was of no other use) to the under side of the floor. Under these circumstances, it was an impossibility for any local current to be *upon* the floor, for warmed air in an apartment condenses most rapidly upon the surrounding walls. Every body who sits in a wall-pew in a church has observed a local current—it being at times so strong downward in a cold day as to produce the sensation of a draft from a broken pane of glass—whereas it is nothing more than a rapid condensation or cooling of the air upon the cold wall or window. Not a particle of this descending sheet of air down the walls *can* come upon the floor, being intercepted by and taken into the aperture by the diminishing of the floor as above stated, at the foot of the walls.

I have merely so far glanced at the manner of the construction of the main floor of a dwelling for the purpose of showing how the cold is kept away from the feet, and consequently the heat of the room more equalized between the head and the feet. The *modus operandi*, or particulars of the construction from beginning to end, will be found explained at large in their proper places.

CHAPTER X.

DRY CLOSETS.

If the City Council of London, some hundreds of years ago, could have foreseen the lamentable picture presented in a late report by the Board of Commissioners to report upon the sanitary condition of the city, appointed in consequence of the effluvia arising from the filthy condition of the Thames, I can not believe that it would ever have permitted the draining of water-closets in the sewers. This great city with its millions of human beings, is in a most perilous and deplorable condition, and if its present system of drainage and sewerage be continued, I can not imagine in what way it can escape depopulation by pestilence.

There should be sewers and there should be drains, no doubt, in every large city, but only to carry off the *water*, not the sordes or the excrementitious matter from the human body. This should all be *carried* away. This idea will at first view be pronounced a most Herculean, as well as an intolerably offensive work. Not so; the residue of twelve ounces of excrement will weigh, when dried, only about two ounces.

Let us suppose, then, a dwelling such as represented in Plates 7, 8, and 9, and the shaft and water-closet as represented on Plate XI., and the house inhabited by a family of ten persons. Then suppose a volume of four feet of air flowing closely over the stone bason made at the bottom to receive the sordes, and up the shaft at the rate of five feet per second all the year round, such will be the power of evaporation that one man will carry upon his back, at one load, the whole of the deposits for years! This surely is no great trouble or expense; nothing to be compared to the expense which is now incurred in keeping the ordinary cesspools and drains in order. In order, however, to make this residuum more conveniently available to be entirely consumed upon the premises where there is even a very small patch for a garden attached, lime, ashes, or plaster of Paris should be thrown down the pipes of the closet, in the proportion of about a gallon per week. This will not only render the mass hard and easily cut up for removal, but the ashes and plaster of Paris will *fix* and retain a great portion of the ammonia, so valuable to flowers, and indeed plants of any kind.

But let us look at the actual state of things as at present. Every water-closet and cesspool is drained into the sewers; the mouths of these sewers are in general run down to the edge of some body of water, which does not always cover the *whole* mouth, as it should do in order to exclude the air, and especially does it not exclude the air at low water where there is a tide. The consequence is, that every house whose drain is not in perfect order becomes a foul-air shaft for the sewer, and the heat and chimneys accelerate the flow of air from the drains, upward and into the building, and especially so when the wind blows into the mouth of the sewer, which it frequently does. The inmates, therefore, of these dwellings have not only to endure the malaria generated within their own dwelling, but have also the *advantage* of that of their neighbors. I have stood at the mouths of many sewers, and instead of experiencing any offensive odor as I had expected, I frequently found a strong draft *into* the sewer. Thus in these cases these sewers carried down the insoluble matter, which, in a sanitary point of view, could do little harm, whilst the noxious gases were carried up into the houses.

These sewers instead of thus becoming the greatest nuisances we have, might in addition to being the conduits for the waste water, be turned to good account in the ventilation of a whole city. Erect foul-air shafts—say about four for every mile—at convenient places, adjacent to the sewers and connected with them by underground ducts, and the exhaustion thus brought to bear upon the sewers, and the sewers upon the drains, would go far to improve the sanitary condition of our cities. If the civic authorities would be at half the expense of the erection of the many furnace-shafts scattered over our largest and most populous places, upon condition of the proprietors allowing a connection with the sewer in the way stated, a very cheap and effectual exhaustion might be had. In general, however, a single shaft erected at or near the mouth of a sewer, would, if properly built, be found sufficient.

CHAPTER XI.

ARCHITECTS AND ARCHITECTURE.

Amid the blaze of light which in this nineteenth century has so illumined the world, architecture alone lies motionless, covered with the dust of ages. Not a single new idea, so far as I know, has been suggested by the profession within the memory of man. Architects, to judge from their productions, appear to think that their sole mission is to ornament the exterior of our buildings. If they succeed in giving us what they call a handsome *façade* or front, and stick on the top and about the building domes, pediments, and pepper-pots to suit, they will rub their hands and chuckle over a competitor as if a great victory had been obtained.

Prof. Youmans, one of the most practical writers of the present day on matters connected with science, uses the following language in relation to our present system of architecture: "There can be little question that the whole policy of warming and ventilating dwellings is yet in an unsettled and transition state; although this affords no apology for neglecting the subject. Much is known, and a great deal may be done about it to promote health and preserve life. Provision should be made for ventilation in the first construction of dwellings, as it may then be effectually and cheaply accomplished. The introduction of adequate arrangements, after the building is finished, is costly and difficult. The necessity is absolute for including ventilating provisions in houses, as well as those for heat. Architects and builders should make them a primary and essential element of their structural design, in accordance with the principles of ventilation as an established art. It is to be regretted that too many in those professions, to which a careless public commits its interest in this particular, are profoundly unconscious of the just claims of the subject, and totally unqualified to deal with it properly. This is hardly a matter of surprise when we recollect how recently it is that science has thrown its light upon the physiological relations of air. It is almost within the memory of men still living that oxygen gas was first *discovered;* and it is within twenty years that Liebig announced the last constant ingredient of the atmosphere. Architecture, on the contrary, rose to the dignity of a regular art thousands of years ago, when men had little more intelligent understanding of the real import of the breathing process than the inferior animals. We have, therefore, little cause for amazement when a book appears upon the subject of architecture, of more than a thousand pages, and dispatches the whole matter of ventilation in ten lines; and that, too, with a sneer. Our buildings are, hence, commonly erected with less reference to healthful comfort than outside show; and ventilation is too much looked upon as a mere matter of tin tubes and knocking out bricks, that may be attended to at any time when it may be thought necessary."

Ordinarily we have the stove-maker, the furnace-maker and the hot-air man, each going his own way; while architects, bricklayers and stone-masons are putting a flue here, a hole there, and a chimney somewhere else; the whole ending in a jumble and confusion worse confounded than the antediluvians in the erection of the Tower of Babel. When, if principles of pure science had been consulted, an uniform system founded upon philosophical principles would have formed part of the original plan.

Proper provision for the ventilation and warming of a building can only be made whilst the building is going up. It must form part of the primary idea. You might as well attempt to put lungs in a full-grown man, and expect them to fulfill their proper function, as to thoroughly ventilate a house after it is built.

I deny that it is the business of the stove-maker or furnace-maker to either warm or ventilate our dwellings. It is the business of the architect alone. Dr. Wyman says: "The architect should design and build, not only with regard to beauty and convenience, but to health and comfort also; this he can not do unless he possesses a clear conception from the beginning of the means by which these several objects are to be accomplished."

The expense of building, with reference to ventilation and warming, will be little or nothing more than by the ordinary mode, excepting the expenditure of *brains* by the architect, and for this he should be well paid; for it will save the proprietor an almost endless expense in future, cutting off as it does all necessity for repairs, and reducing the consumption of fuel to an almost indefinite extent.

In Europe, especially in the southern parts, whose mode of building has descended down to and is copied by our architects, their main object in the ventilation of their buildings is to *cool* them, and hence their high ceilings and open stairways; but in the northern parts of this continent, where we require artificial warmth for much the greater part of the year, our object should be the *warming* of our dwellings; and it is strange that our architects have not long ago seen that the accomplishment of the latter object requires a course of proceeding exactly the reverse of the other. What we now want and must have are architects *for a cold climate.* We want a few bold men of this noble profession, who can *afford* to stem the force of prejudice against the low ceilings and close halls which the people have been taught by their predecessors to regard as vulgar. I lay it down as a postulate that no dwelling-house in this country can be made comfortable, as regards ventilation and warmth in winter, with high ceilings and open stairways, no matter what quantity of fuel the inmates may consume.

The provision of fuel is getting to be felt as a heavy burthen upon the housekeeper; and when I state, after sixteen years' experimenting in the warming and ventilation of buildings in Canada, that it takes considerably more than double the quantity of fuel to warm a room twelve feet high that it does to warm one nine feet high, of the same ground area, it behooves every man to pause before he sacrifices so much money and comfort to a mere prejudice.

We must have close halls. I never yet could see the utility of an open staircase in the hall of a dwelling; certainly in a cold country it is a most mischievous arrangement. We all know it as a fact, that heat our houses as we will, the first story, which should be the warmest, is always the coldest when there is an open stairway; and if, in a very cold day we bring the temperature below up to the requisite point, the upper story will be too hot. With a close hall we can easily make the first story warmer than the upper one, and equalize the temperature at pleasure. This will save a great deal of fuel, and in the course of even one winter add much comfort to the inmates of such a dwelling.

People do not reflect, when arranging to warm their houses, that warm air is lighter than cold; and that if they will only give it a chance it will rise out of the top of the house, giving place to cold air, which rushes in through every open door, and finds its way through every crack and crevice which may have been left by the carelessness of the builder. This is the fact, however; and the open staircase furnishes a passage for the escape of the warmed air into rooms that are seldom used during the day, except in large cities; and there, if the lower story is comfortable, the upper ones are much too warm. It is truly marvelous that our architects will still persist in recommending the open stairway; and, more marvelous still, that proprietors should submit to it.

We must have no rooms over ten feet high, and nine feet would be better, except in the case of churches and theatres, where special provision is made for their warming. I know there will be strong opposition to any such proposition as the above; for hitherto high ceilings have been looked upon as a sort of substitute for ventilation, and as the need of ventilation was felt more in cities than in the open country, the ceilings were made higher, and from that it has grown into the fashion. But if thorough ventilation can be secured, the great advantage a low ceiling possesses over a high one in the saving of fuel, as well as in the rapidity of a change of air, will I think induce those people who are governed by common-sense, to adopt it. That this thorough ventilation can be secured, there is no shadow of a doubt, and with high ceilings too, if people choose, but they will find that their fuel-bills will run up just in proportion as they increase the height of their rooms. For all public buildings, which are only occasionally used, and are then likely to contain large assemblages of people, from whom a considerable quantity of animal heat must flow, the ceilings may be high, and indeed should be, for there the provision for warming becomes a minor consideration.

We must provide our dwellings with double windows. It will be seen by the following quotations, how pre-eminently important this is in our northern climate, where, for a large portion of the year, such a great difference exists between the temperature of the inside and that of the outside air.

Dr. Wyman says: "When a cold window makes a part of one of the walls, a constant current of cold air descends along it, which is often mistaken for that which enters the window from without; but it will exist without that, and can not be prevented by closely-fitted sashes or any care in caulking their crevices. The unpleasant effect of this fall of air from a number of large windows, as in churches, and their great influence in lowering the temperature of the room, is much greater than is usually supposed, especially in buildings heated by warmed air, when the walls do not feel the influence of radiated heat. In our New-England climate, where the temperature not unfrequently approaches zero, and is often below the freezing-point, there would be a vast saving of heat, if our churches, court-rooms, and other public buildings could be preserved from this cooling process. This can be done by means of double windows, fitting closely and inclosing between them a quantity of air. Air, as is well known, transmits heat only by a change of position among its particles; each particle may receive a portion of heat from a heated body, and, by coming in contact with another less heated body, communicate its heat to it, but not otherwise. One particle never communicates its heat to another particle. Hence if glass or any other material which transmits light be placed at two or three inches' distance from the glass, the inner sash will be kept warm, the circulation of the air between the sashes going on so slowly; consequently less heat will escape from the room."

Dr. Reid remarks: "A window is always a source of descending currents, which take place altogether independently of any influx of cold air. Even where the window has been rendered absolutely air-tight, it is obvious that if the glass be maintained at a low temperature by the external atmosphere, the air in contact with it must become cold and dense, falling accordingly, and producing a continuously descending current so long as the glass is colder than the rest of the apartment. These currents are a common source of discomfort and disease, particularly rheumatisms, colds, and inflammations, sometimes terminating in death. The cause of complaint is more frequent in public buildings and in all situations where large windows are introduced."

It is plainly seen from the foregoing that health is not only promoted by double windows, but that a great saving of fuel is effected also. Certainly no stronger inducements can be held out for their adoption.

EXPLANATION OF THE PLATES.

PLATE I. FIGURES 1 AND 2.

THE WAY TO LAY THE LOWER OR MAIN FLOOR OF A HOUSE.

A, Joists; B, firring over joists Fig. 1, and under joists Fig. 2; C, two-inch space left between the floor and the wall of apartment, to be covered from sight by the plinth of an open iron base, A, Plate III. Fig. 1; D, floor; E, lath and plaster of ceiling of basement; F, plastering of wall; G, wood-moulding of top of base, and to which top of iron base is screwed, the bottom being screwed to the floor. See Plate III. Fig. 1.

It is only the main or first floor of a building or apartment that the air is required to circulate *under* the floor, the object being to prevent all local currents over the *top*, and to warm the floor in order to more equalize the temperature between the head and the feet. To exhaust the air of an *upper* room, *under* the floor of an upper room, would do more harm than good; because the greatest heat lies against the ceiling of the room below; and consequently the air between these *overhead* joists is always warm; and consequently its exhaustion would rather lower than raise the temperature.

These two diagrams show how the air may be made to circulate under the floor, and gather to any one point on any side of the room with which the chimney or foul-air shaft may happen to be connected; it follows then, of course, that the air must of necessity flow *across* as well as lengthwise of the joists. To make room, therefore, for this cross-circulation, I lay firring, B, across the joists, either on the *top*, as in Fig. 1, or on the bottom, as in Fig. 2; this will leave a space, according to the size of the firring, either at the top or bottom of the joists for this cross-circulation. Two-inch firring for any ordinary-sized room will be sufficient; but for churches, theatres, or other very large apartments, four or even six-inch firring may be required. You perceive the space C, between the wall of the room and the edge of the floor, two inches. The arrows show the air falling down into this space, all around the room, except of course where the doors and hearths, etc., come.

PLATE II.

SHOWING HOW THE CHIMNEY MAY BE CONSTRUCTED SO AS TO EXHAUST THE AIR EITHER ABOVE OR BELOW THE FLOOR AT PLEASURE, AND HOW TO WARM AND VENTILATE OLD BUILDINGS WHERE NO PROVISION FOR DRAWING THE AIR UNDER THE FLOOR EXISTS.

A, aperture in basement, in foul-air flue; B, sliding-valve to regulate quantity of air; C, joists; D, firring; E, open base; F, two-inch space between floor and wall; G, sliding-valve in chimney-board, more particularly to be used in old buildings; H, fire-place or grate-flue. The size of both these flues, in ordinary dwellings, are either of them intended to exhaust a room say twenty feet square, and an adjoining bed-room or two, and should measure about one hundred and forty-four inches in the clear, each. [In all cases, where I speak of a flue or duct containing so many inches, I mean that a cross-section of the flue or duct shall measure that; that is, that on any rectangular flue or duct, one side multiplied into the other shall make that number of inches. For example: a one-hundred-and-forty-four-inch flue may be made $12 \times 12 = 144$, $16 \times 9 = 144$, $18 \times 8 = 144$, $24 \times 6 = 144$; and so on.] A circle is the best shape for any flue; a square flue is the next best; a 16×9 is better than a 24×6 flue, and so on; because the sum of all the sides is less as they approach to equality, and therefore there will be less friction. These two flues, A, H, should be kept separate from each other until you get to within about two feet of the ceiling of the room of the last or highest story, where they must be brought together as seen in Plate LI., (the dotted circle intended to represent the place where your smoke-pipe is to go in, if you have one.) From thence you carry out to the top of the chimney only one flue, which, assuming that you start at the bottom with two twelve-inch-square flues, will be only two hundred and sixteen inches or one and a half feet. If you can persuade the proprietor to make the partition or division between the two flues, of sheet-iron instead of brick, so much the better for him; because, in that case, whenever there is a fire in the grate or fireplace, it is almost as good as a fire in the ventilating flue itself. I, floors; K, basement; L, handle of regulating-valve. The aperture A may be made in the front or breast of the chimney in the basement, as well as in the side, as here shown, if more convenient. The valve B (to which you see the handle is riveted) I make of sheet-iron, as well as the rebate (rabbit) in which it slides. If the flue is 12 x 12, I make the aperture in the wall, 14 x 14 inches. I then take a sheet of Russia-iron and cut an aperture out of the middle, 14 x 14 inches, rivet on the rebate, see that the wall is brought to a perfect face, and out of wind, (by mortar if necessary;) put the sheet to its proper place, and nail it firmly on. This, now, in the bottom of a chimney forty feet high, or in one which will always during cold weather have fire in one of the flues, will do. But as it may happen, that from the carelessness or ignorance of your workmen, the flues of your chimney may have been badly constructed, not smoothly plastered, or from there being no chimney-cap, as represented on Plates XXVI. or XXX.; or from the chimney being a very short one; or from the peculiar locality of the house; or from there being no plenum fresh-air duct as represented by Plates XXXI. or XXXII.; or from there being another chimney or chimneys in the same or adjoining room, open to this room, with strong fires in them when there is no fire in this; that there may happen to be a *back-draft* in this ventilating flue. In order to prevent inconvenience from this cause, I hang a self-acting valve entirely *within* the aperture A, to open inward and work horizontally. I make this door of light tin, and hang it perfectly plumb, with very loose and easy-going hinges. This Plate being necessarily drawn on so small a scale, this valve could not be intelligibly represented here; but if you will turn to Plate XLIV., and look at Y, you will at once understand what I mean. This valve Y must have a "stop" put behind it, for the same reason as that given in the reference to O, on that Plate.

The throat of every fireplace-flue, and of every grate-flue, as well as every ventilating flue in every dwelling, must be furnished with means by which they may be closed and opened at pleasure; so that when, in cold weather, the particular apartments with which they may be connected may not be required to be warmed, the flues may be closed and thus prevent a waste of fuel. This may be so easily effected by grate-makers and chimney-builders, that no farther observations are required from me.

PLATE III. Figures 1, 2, and 3.

OPEN BASE.

A, open base. This perforated plinth of a base I have made of iron, in pieces thirty inches in length, and so cast as to allow the top edge to be screwed to the top moulding, B, and the lower edge to the floor, C. It may, however, be made of wood, and the perforations made with a "bit" or auger, if preferred. D, joists, Figs. 1 and 2. E, firring laid across them to allow the air to circulate *crosswise* of the joists, Figs. 1 and 2. F, Fig. 3, aperture made through the wall to allow the air to enter the apartment to be ventilated. It is made as near the ceiling as possible, so that, in cold weather, the *warmest* air may be admitted. The aperture is covered by a fan-register, as you see, to be opened and closed by means of cords hanging down within reach. This is a more economical arrangement to get the air into a room than by the open cornice, as seen in Plate XXXIII. Fig. 1.

PLATE IV.

"FOUL-AIR-GATHERING DUCT."

A, Ventilating flue. B, regulating valve. C, foul-air-gathering duct or box. D, joists. E, firring. F, hearth. G, floor. H, top moulding of base.

Before I enter into the particulars of this foul-air-gathering duct, it is necessary for me to explain a very important point connected with the general subject, and unless this be thoroughly understood and acted upon throughout from the very commencement to the very end of the building, all attempts to warm it will be futile.

Before you begin a building to be ventilated and warmed, you calculate the *quantity* of fresh air you require to do the work. We will assume that a dwelling forty-feet square, and two or even three stories high, and the rooms nine or ten feet between joists, will require four feet of fresh air, flowing at the rate of five feet per second, (and this may be taken as an approximation to the quantity actually required for such a building,) and you have four good chimneys with a ventilating flue of one foot in each, as you see in this figure at A; you will then construct your fresh-air duct accordingly, and will bring in a quantity of air 4 x 5 = 20 x 60 = 1200 cubic feet of air per minute. It may then fairly be supposed that your four foul-air flues will take out of the building, the same quantity that your fresh-air duct will bring under your air-warmer; for although the air will expand thirty or forty or perhaps fifty per cent, yet by the cooling effects of the walls, etc., of the house which it will wash, it will again shrink in bulk as the temperature is diminished. We will therefore suppose that your foul-air flues will carry out of your building the same quantity that your fresh-air duct will bring into it, *but not a particle more.* It must be obvious then to any one that, if any *other* air than that which has come through your air-warmer, and has there been warmed, comes in at any one point between the air-warmer and the top of your chimney or shaft, it will prevent just that quantity from coming into the fresh-air duct, and consequently from being warmed. For example, if four feet of the external air should be let into the basement of this building through a door or window, and the whole of the four flues A be unprotected against it, not a particle of air would come through the air-warmer, all circulation through the rooms would cease in an instant, and, of course, your house would become as cold as if you had no fire within it.

You now see the importance of this foul-air-gathering duct or box—it is to keep the flue A protected against the influx of the external air. I give it this name merely for the purpose of reference, because into it is "gathered" all the air which is rapidly circulating amongst the joists of the room above, (as you will perceive by the arrows,) and conducts it into the shaft or chimney, free from any mixture of external or foreign air. The duct or box is cut away in the drawing merely to let you see its object, but it must be made air-tight, and may be constructed of wood, brick, or any other suitable material. I have also cut away the flooring and the firring and joists, so that you can see into the top of it.

The flue A may be made in the front or breast of the chimney-foundation as well as at the side where I have placed it; except for the convenience of working the valve, the handle of which you see goes up into the room above in a convenient place for handling it.

Again, in order to impress it more fully upon your mind:

By the two-inch aperture around the room, and by the firring on the top of the joists which we have already described, you see that provision is made for the warmed air to be drawn *under* the floor, and also for its free circulation, after it is got there, to any one point required. All this, however, will be of no avail unless we can connect the whole body of air lying in the space between the floor and the ceiling of the room or basement below, with the chimney or foul-air flue. The box or duct C is for this purpose.

A building to be warmed and ventilated in a cold climate must be regarded as the middle part of *one air-tight tube*, swelled in the centre to the dimension of a house, but nevertheless strictly a part of the tube. The one end of this tube you see at B, Plate X., the other end being the top of the shaft, Plate XXVI. or Plate XXX. Now, *in the building for ventilation this idea must be constantly kept in mind.* There must be no break in this tube from one end to the other—it must remain one connected, uninterrupted *tube*. It is true that the opening of *outer* doors will somewhat interfere with this idea, but as this is but occasional, and as it is inevitable, we must put up with it; but in the basement of a building the outer door of which is frequently left open for hours and perhaps days, the whole thing may be rendered nugatory; the *outer* air filling the ventilating flue instead of that which comes through the air-warmer, the ventilation as well as the warming would instantly cease, as much so as if you cut a water tube in two. You will now see the use of this box or duct. It answers the same purpose here that a water-duct does in supplying a mill where the water for its supply has to be carried through a river or some other body of water. In the plate it is broken away at the bottom in order to show its connection with the chimney-flue A, as also the floor, firring, and joists at the top are broken off to show how the air from between the joists falls into it as indicated by the arrows at C. This duct may be made to carry the air of the room to any *distant* shaft, supposing there was none here, either within or even out of the building if necessary; it may lie horizontally and be carried to any distance, only it must be kept air-tight, every inch of it. I have been particular in the description of this part of the plate, as it is an important part of ventilation, and also as it will render unnecessary any lengthy remarks upon Plates V. and VI.

The references at the head of this description will explain the rest.

PLATE V.

ANOTHER VIEW OF THE FOUL-AIR-GATHERING DUCT.

A, foul-air flue. B, sliding valve to regulate quantity of air. C, box or duct to connect the foul air circulating among the joists above with the chimney or shaft. D, joists. E, firring. F, under-side of flooring of the room above, as also top of basement floor. G, plastering of ceiling of basement. H, wall of basement and foundation of chimney.

This plate is made to be viewed from the basement, in order that you may see that the top of the front side of the duct must be cut off even with *the bottoms of the joists*. This duct or box must be made broad enough to comprehend as many *spaces* between the joists as will be equal to carry the full quantity of air that comes into the room above. In this plate it is broad enough to take in two spaces, which, as joists are, I believe, never less than one foot apart, will carry two feet of air. The draft of air will then be from all sides of the room into these two spaces, along which spaces the air will flow, (as indicated by the arrows,) and fall into the top of the duct or box, and so up the chimney-flue A.

PLATE VI.

A THIRD VIEW OF THE FOUL-AIR-GATHERING DUCT.

A, mouth of ventilating flue. B, sliding regulating valve. C, box or duct made to receive the foul air from the space between the main floor, F, of the room above and the ceiling, G, of the basement, and convey it to the foul-air flue A. D, joists. E, firring. F, under-side of the flooring of the room above. G, plastering of the basement ceiling. H, H, wall of basement and foundation of chimney.

The only difference between this diagram and Plate V., is that this shows that the plastering G of the basement must be made air-tight around the *top* of the duct or box.

I may here, as well as any where else, remark that a room or rooms may be thus ventilated without any chimney being within or near it. All you have to do is to have a good ventilating-flue *somewhere*, and by means of a wooden duct of sufficient size, connect the foul air of the room with it. The foul-air shaft, or chimney, will do its work at one hundred feet distant as efficiently as if it adjoined upon the room.

PLATES VII., VIII., AND IX.

ELEVATION AND PLAN OF DWELLING.

The references on these several plates themselves are sufficiently explicit.

I give the elevation and plan of a dwelling-house, not so much because there is any thing new in them, as to give me an opportunity, when viewed in connection with other plates, and especially Plate X., of explaining how a building may be ventilated in the most economical and efficient manner.

I have elsewhere given my opinion as to the propriety of having cellars under a dwelling-house. There is none intended to be under this. The bottoms of the joists are to lie within a few inches of the ground, leaving just room enough for the air to circulate crosswise under them, except under the hall, where the ground is to be taken away from one to two feet below the bottoms of the joists. The hall is to be ten feet wide, and run back under the water-closets, up to the foul-air shaft, which stands in the kitchen.

For further explanations in regard to the ventilation of this building, see observations on Plate X., being a perspective view of the foundation.

PLATES X. AND XI.

FOUNDATION OF A DWELLING-HOUSE, BY WHICH THE PRACTICAL WORKING OF THE VENTILATION MAY BE ILLUSTRATED.

A, foul-air shaft, or chimney. B, fresh-air duct. C, air-warmer. D, open base. E, ground or pavement under the hall, (or bottom of common foul-air chamber,) and which may be regarded as the horizontal part of foul-air shaft. F, apertures, with sliding valves to close them, to let the air lying under the respective rooms flow into the common foul-air chamber. G, door to obtain access to the deposits from the water-closets. H, H, pipes to convey deposits down to the receptacle at foot of shaft, (in Plate XI. these pipes are designated by D.) I, I, arched entrances of foul air into shaft. K, jambs of kitchen fire-place.

I have caused all my plates to be drawn upon as large a scale as the size of my book will allow; it necessarily follows, therefore, that the several *parts* of the building must be on different "scales." The whole of the Plates VII., VIII., IX., X., XI., XXVI, or XXX., belong to this building of which we are now about to describe the practical working of the ventilation.

I have endeavored to give, in this plate, a perspective view of the foundation and lower part of the dwelling, up to the height of the foundation walls, under the level of the floor, (some parts a little above it.)

The building has no cellar or basement story under it; the lower edge of the joists should lie but a few (say six) inches above the ground or pavement, except the joists supporting the *hall*-floor, where the ground or pavement should be about two feet below the joists. When the hall-floor is fully laid, air-tight from front to rear, up to the foul-air shaft, as seen about where the air-warmer C stands, the space below it will then be one continuous air-tight duct ten feet wide and upward of forty feet long, and may be regarded as the first or front end of the foul-air shaft itself. All the foundation or dwarf-walls, where the partitions above are to be of wood, must be sure to be worked up air-tight *to the floor;* all the spaces on top of the walls, *between the joists*, to the top of them, must be filled air-tight, so that when the floors come to be laid, not a particle of air can escape from the spaces under the several apartments, into this space under the hall at this point. This part of the work has hitherto been considered of little importance, and has been so slurred over in consequence, that it will require the most unremitting attention of the overseer of the work to insure its being done. You will note here that in these precautions I am only anticipating the case where the partitions *above* the foundation-walls are to be made of *wood;* of course if the foundation-walls are continued *up* throughout with stone or brick, the spaces between the ends of the joists must of necessity be air-tight. The foundation-walls, of course, divide the principal apartments *below* the floor the same as they are divided *above* the floor.

The foul-air shaft, or chimney, should not be less than forty feet in height — the higher the better; nor must the flue within it measure less than six nor more than eight feet, eight hundred and sixty-four or eleven hundred and fifty-two inches. This shaft may be erected at any convenient point, so that it be connected with what I may now call the horizontal part of it, namely, the space or duct under the hall; but on account of having the advantage of the kitchen-fire, as also of the smoke-pipe of the air-warmer, which may be run into it along the top of the hall in one of the upper stories, as well as for the convenient location of the water-closets, I prefer to place it here. Supposing now that this shaft will do its duty, all air which may be let into the horizontal part of it will be immediately drawn toward and up the perpendicular part of it.

The floors of the several rooms, in this as well as in all other cases where there is no basement or cellar under the building, may be laid directly upon the joists, there being no need of firring or of a "foul-air-gathering duct;" the circulation every way being uninterrupted the instant it falls through the open base, and gets under the floor to any point where the aperture for its escape may be made. Indeed, with this two-inch open space, as seen at C, Plates I. and II., Figs. 1 and 2, the whole body of air, from the ceiling of the room *down to the ground or pavement*, is to all intents and purposes one body, as much so as if there were no floor at all in the room; and may, therefore, be as readily drawn off *below* the floor as *above* it.

The *hall-floor*, which forms the top of the horizontal part of the foul-air shaft, must not have an open base, but must be laid as usual, and kept air-tight up to the side-walls throughout.

The foul-air shaft, or chimney, is much the most important part of this building; as the wind-pipe in an animal is the chief organ of the breathing process, so is this structure the main feature in the lungs of a building; and upon its working depend both the ventilation and warming. And in order to make its operation more plain, it may be well for us to imagine that Plates XI. and XXVI., or XXX., being drawn on a larger scale, be put together and placed where this shaft, A, stands. We have now, as before stated, one continuous air-tight shaft, of from eighty to one hundred feet long; but its efficiency depends almost wholly upon the *height* of the perpendicular part.

In order now to open up a communication between the outer air at B, and the top of the shaft, and without which no air can come into or go out of the building, we must imagine the four apertures (one under each of the principal rooms) covered by the four sliding valves, F, made.

These should, of course, be made when the walls are going up, by setting in wooden frames, each of which must allow two feet of air to pass through from under each of the rooms, into this space, E, under the hall, whenever the valves, F, are raised; which is done by means of the handles, which, as you see, run up into the rooms. Nor is the communication yet complete until another necessary part of the lungs are put in. These are the fan-registers, F, Plate III., Fig. 3. These must be put into the wall as high up and as near the ceiling as possible, (in order to get the hottest air you have into the apartments,) and of a size to allow not less than one hundred and forty-four inches of air to pass through from the hall into the apartments to be warmed and ventilated.

For the manner of bringing the outside or fresh air through the air-warmer, and into the building, I must refer you to the description of Plates XXXI. and XXXII. In the mean time, we will suppose the duct, B, under the drawing-room, to be one branch of the plenum fresh-air duct; that the hall-floor is cut through the proper size, (which, if my air-warmer, Plate LII., is used, will be 40 x 21 inches for the No. 2 and 29 x 16 inches for the No. 1; see Fig. 4;) that the fresh-air duct is connected, air-tight, with the under side of the aperture, cut through the floor; that the air-warmer, C, is placed over it; that the "fresh-air regulator," Fig. 3, is opened; that the fan-registers, F, Fig. 3, Plate III., be opened; that the apertures, F, leading from under the apartments into the common foul-air chamber, E, are all open; we shall then have a full and free flow of air throughout every apartment of the lower or first story, and out of the top of the shaft.

In regard to the upper story or stories of a building, a register or registers either in the form of Fig. 2, Plate XXXIII., or any other form, may be put in the ceiling of the lower hall, or rather into the floor of the upper hall, so as to allow a small quantity of the fresh air to ascend. For economy *one* of these must be put in, and the smoke-pipe of the air-warmer go through it and into the bottom of a dumb-stove, D, Figs. 1 and 2, Plates XII. and XIII., which is a necessary adjunct for the warming of every upper story, and from the top of the dumb stove into the foul-air shaft, or if you have no foul-air shaft, across one of the rooms into one of the chimneys, as seen on Plates XII. and XIII., Figs. 1 and 2. If you have an open stairway, (to which I am altogether opposed, for reasons given in the chapter on Architects and Architecture,) the exhaustion will take place down the stairway in common with that of the lower rooms, or in case of a close hall, you exhaust by opening an aperture near the floor into the chimney, and into which a register is put.

The exhaustion of one foot of air will be sufficient for any upper story; or you may cut off three inches of the bottom of your stair-door, and the exhaustion will take place as in the case of an open stairway. If the air is warmed in the *basement*, the whole body, so warmed, is let up through a register of sufficient size at once into the body of the hall, and the ventilation and warming goes on just in the same way as above described.

There must be fan-registers put into the upper-story rooms as in the lower story, but no open base—the exhaustion taking place *above* the floor, for the reasons elsewhere given. The warming of the lower story will be facilitated by placing the fan-registers as near the air-warmer, or source of heat, as circumstances will admit.

No dwelling in this climate should be without a good, substantial, air-tight portico or vestibule—the doors of which should open outward and be hung on the same sides, with good springs upon them, and these as far from each other as circumstances will admit; the outer door should open *against* the prevailing winter wind.

We will now suppose that it is a cold winter's day, our dwelling finished "to the turn of the key," every valve and fan-register open, and that four feet of air is flowing through it at the rate of five feet a second, equal to 1200 feet per minute; that the building is 50 x 40 on the ground, and twenty feet high, equal to 40,000 feet; that the air-warmer has a good, sharp, bright winter's fire of coal or wood in it. It is evident that in little more than half an hour the whole of the old air will have gone out and have been succeeded by warmed fresh air from the outside. This is a very close approximation to what will take place in every instance under like circumstances.

It may be that some of the apartments may not be needed; in such case the valves, F, under such apartments may be left closed, and no warmed air can, so long as these valves remain closed, enter into such rooms, and consequently a considerable quantity of fuel may thus be saved.

PLATE XI.

DRY CLOSETS, ATTACHMENT TO FOUL-AIR SHAFT.

A, shaft; B, seat of closet in first story; C, seat of closet in second story; D D, pipes to convey the sordes down into the basin at bottom of shaft; E E, floors of closets; F, joists; G, the ground; H, stone or cast-metal basin to receive the sordes.

I have caused this part of the foul-air shaft to be drawn on a larger scale than that on Plate X., in order that the several parts may be rendered as intelligible as possible. The pipes, D, will be best made to increase in size from where they join the top flange (where they should be about five inches in diameter) down to the lower end.

For the utility of this appendage to the foul-air shaft, I beg to refer the reader to the chapter on "Dry Closets."

Under a proper system of full ventilation, it is of course intended that every dwelling-house in this cold climate should be built air-tight, but we are all aware that no human skill is equal to the accomplishment of it, much less can we expect such a thing from the present careless and slovenly mode of building. No air should be allowed to enter the building except *through the channel purposely provided for it*, so that it may first be warmed; all the air, therefore, which comes in through the "cracks and crevices," which number by the million upon millions, through the solid material of the walls, as well as where these materials are joined together, is constantly adding to the coldness of the apartments, and this coldness increases in proportion to the power of your exhaustion *out* of them by means of chimneys. If you have no chimneys, little or no air can come in, either from cracks or crevices or any other quarter. It is important, therefore, that whilst we must have sufficient exhaustion within the building, in order to draw the quantity of air through the air-warmer or warming-machine necessary to warm the house, we should not have *any more*.

In order to show us the vast quantity of air which enters a house from *cracks and crevices alone*, we have only to go to the top of a chimney, in what we call a well-built house, closely shut up, when you will perceive, in some cases, especially in a windy day, an out-flow sufficiently strong to blow a man's hat off! This is *all* from the cracks and crevices of the building. It follows, therefore, that if we perfectly close the top of this chimney, or close it at the bottom, which will amount to the same thing, little or no air can come into the building through the cracks and crevices.

Again, if a pane of glass be broken out of the window of a room having a good chimney in it, the external air will rush in with great velocity; but if you open a whole window or a door, which will let in as much as, or more than the chimney-flue will take up, the velocity, and consequently the quantity coming through the broken pane will much slacken if not altogether cease — the room becoming surcharged by the larger quantity, fills the flue to the exclusion of the air admitted by the smaller aperture. By the same rule, if we admit as large a quantity of warmed air as we are contemplating to be necessary to warm a building, very little, if any, of the external cold air will be drawn through the cracks and crevices. In a very cold night the valve in the fresh-air duct, B, Plate X., must of course be entirely closed on going to bed, the air-warmer acting for the remainder of the night merely as a common stove, (though in an ordinary winter night, and a good hard-coal fire, a small quantity may be left on,) the whole building being then warm and full of pure air, there is no necessity to introduce any more until morning. Now, suppose in this case the *whole* of the aperture, I I, be left open all night, it is evident that the whole power of the suction or draft operating upon the cracks and crevices alone will make the house cold before morning; whereas if the draft up the shaft be closed, the greater portion of the warmth will be retained.

I recommend, therefore, that a good sliding-valve upon the principle of that seen at H, Fig. 2, Plate XVI., be made at the foot of the shaft; so that during very cold nights the power of the draft may be checked, or entirely closed. If this "closet" attachment be adopted, it will be easy to put stops near the bottom so as to allow space enough left for the sordes; or the arch, I I, may be extended a little out from the shaft, and the valve closed beyond the sordes. A building thus warmed by air, and the suction or draft of the foul-air shaft (or common chimneys if no foul-air shaft) be thus checked, will keep warm for a much longer period of time than by allowing the full draft of the chimney to operate upon the cracks and crevices.

Once more: In an intensely cold day our warming-machine may not be equal to the work of warming so large a quantity of air as the *whole* opening in the bottom of the foul-air shaft may draw through the air-warming machine. If, in such an emergency, we close the valves of the *fresh-air* duct, B, Plate X., and *thus* diminish the quantity of air, this will but increase the suction upon the cracks and crevices; whereas, if we diminish this draft *at the foul-air shaft*, by partly closing the aperture, it will lessen the quantity of air coming through the air-warmer as effectually as if it were lessened by partly closing the fresh-air duct valve; and will at the same time lessen the quantity coming through the cracks and crevices; whilst by the other course it will *increase* the influx through the cracks and crevices.

PLATES XII. AND XIII.

VENTILATION OF OLD BUILDINGS.

A, foul-air shafts or chimneys. B, fresh-air ducts. C, air-warmers. D, dumb-stoves. E, apertures through walls close to the ceiling, to admit warmed air into the apartments to be warmed.

By old buildings I mean all habitations not expressly built for ventilation.

Hitherto I have purposely avoided allusions to my own air-warmers except where it could not be helped, lest I might lay myself open to the ungenerous construction of "puffing my own wares." I am, however, now that I am considering the subject of the warming and ventilation of old buildings, obliged to refer to these machines, because people will no longer incur the expense of furnaces set in the cellar or basement for the purpose of heating perhaps an old and dilapidated dwelling or a cottage or school-house, even though such furnaces would do the work as efficiently and with the same economy of fuel and safety to health. Moreover there being no other portable air-warmer extant—at least I have never seen or heard of one—intended to warm a building by means of air, it would be impossible for me to do otherwise than to take my own invention for the purpose of reference.

Of course the exhaustion of the air out of an old building must take place *above* the floor. The change of air within such a house may be as rapid and effectual as in a house built for ventilation, but the same degree of comfort and economy of fuel need not be expected.

The first requisite toward the proper warming and ventilation of an old building is plenty of chimneys, at least one in every principal room. The second thing to examine is the height of the rooms on the first or main floor; if they are more than ten or at most twelve feet high, I advise you not to undertake its warming in this climate. The third thing required is a wide hall and room for the air-warmer to stand; and the fourth is to see that, if in country places, you can get the fresh air from that side of the building which is most exposed to the prevailing winds in winter. If you find all these essentials to your mind, the rest, after undergoing the operation once, will be comparatively an easy matter to you.

The air-warmer should stand as near to its work as possible, which in general will be in the centre of the hall. If, however, the hall be too narrow for this, it may be set on one side, and may stand within six inches and even within three or four inches of the wall, if necessary, without any danger, as it has a double casing, between which the air is rapidly circulating, and thus keeps the outside plate comparatively cool. Having decided upon the place and the sized air-warmer you are to use, you cut the aperture through the hall-floor—if for the smaller, No. 1, 29 x 16 inches, if for the No. 2, the aperture must be cut 41 x 21 inches. I have now got up two other kinds of hall-stoves, (see Plates LIII. and LIV.,) which are perhaps as efficient as those seen on Plate LII. These apertures need be only to let through, for the No. 3, 120 inches; for the No. 5, 200 inches of air.

By many experiments, I have ascertained that the No. 2 air-warmer is equal to the warming of a dwelling forty feet square, and two or three stories high, provided the lower rooms are not more than ten feet high. The No. 1 will be sufficient for a dwelling thirty feet square, and two stories high, and rooms no higher than ten feet.

Although there are iron registers (see Plate LII.) under the air-warmers, by which the quantity of air may be regulated, yet I recommend also a good air-tight valve to be placed at some convenient place in the fresh-air duct, for use in winter nights.

By reference to the drawings you will see two different kinds of "fresh-air duct," B, Plates XII. and XIII., and Plates XXXI. and XXXII. These need very little explanation beyond giving their size. If you use the No. 1 stove, you require not less than 300 inches; if the No. 2, not less than 500 inches of air. It will be safer, in both cases, in order to make sure work, to add fifty inches, if convenient.

If, in order to get this sized air-duct under the air-warmer, you are from some cause obliged to make *turns* in it, let these turns be gradual, not square or right-angled turns.

You will see by the plates that the cellar or basement-wall is opened, and the duct made either of wood as in Plate XIII., or of brick, as in Plate XII., or, which is better still, as in Plates XXXI. or XXXII., or Plate LIV., which is best of all, and connects with the aperture cut through the floor, and over which the air-warmer is to be set as seen in the plates. It is *all*-important that this fresh-air duct should be *perfectly air-tight*, that not a particle of air should escape out of it, except that which comes up through the air-warmer. If, therefore, you make it of wood, it must be of the best-seasoned boards or planks; these must be well matched and put together with paint, and, if convenient, as a further security, paste or glue some good, tight fabric or felting over the joints. The nicest and most difficult thing you will find to be to join this duct air-tight around the aperture *under the floor and joists*, but it must be so done. If you make the duct of brick, you follow the course indicated by the drawings and descriptions.

The top of the horizontal part of this brick duct, whether laid *above* the cellar-floor or *under* it as in the plate, ought, of course, to be an arch, and the wall of this arch not less than eight inches or the *length* of a brick *thick*. If laid on the top of the cellar-floor, it must be protected by a plank covering. The valve, for shutting off the air at night, will be best placed in one of the perpendicular or sloping ends of the duct.

This brick duct will be more easily made air-tight around the under-side of the aperture through the hall-floor than the wooden one; and I should recommend, where room can be spared in the basement for the structure, that you should, before you put up a wooden duct, make a brick erection from the cellar-floor, perpendicularly up and around the aperture, making the inside exactly the same size as the aperture through the hall floor, leaving an opening in the side of this erection to put the end or ends of the wooden duct in, and have it well plastered inside. (See Plate LIV.) The wooden duct may be hung to the joists, as more particularly indicated in Plate XXXI.

The next thing to be done is to provide for letting the warmed air from the hall into the several apartments. You make an aperture as close to the ceiling or cornice as you can into each room. (See E, Plates XII. and XIII., and also F, Fig. 3, Plate III.)

Make a frame of inch stuff, as broad as the wall is thick, so that the frame will be flush with both sides of the wall. For all lower rooms not over twenty feet square, and ten feet high, the frame should not be less in size than will pass 200 inches of air through them; those for upper rooms need not be larger than 144 inches. Into these frames put fan-registers, so constructed as to open and close by cords. These registers must be put on the *room*-side of the wall, not on the hall-side. (See F, Plate III., Fig. 3.) Any other shaped register may be used.

Then make a good tight-fitting chimney-board for every fire-place in the house, cutting out of or near the bottom an aperture at least equal to 288 inches in size, and over this make a sliding-valve to regulate the quantity of air, or close it entirely, at pleasure, G, Plate II.

The next is a good tight portico. To this, the proprietor, if he is a sensible man, will not object, only tell him that it will not only save him a very considerable quantity of fuel, but will wonderfully add to the comfort of his family. If the hall will admit of it, this convenient appendage may be made inside by throwing a partition across the hall.

The *outside* or portico-door must be hung so as to open *outward*, and on that side against which the prevailing winds blow in winter. For example, in Canada the prevailing winds in winter blow from the north and west, so that if a house faces the south, I hang my *outside* door on the west side; and so, whatever aspect the front of the building may have, hang your portico or outside door so that it will in all cases open *toward* the prevailing quarter.

All doors in every building, but especially those in a dwelling-house, should have good springs upon them; but one for the portico-door is indispensable.

Then in all houses having open staircases there is another difficulty to overcome. A large proportion of the warmest air will tend to flow up the stairway, leaving that which is of a lower temperature for the lower story; and to add to this untoward state of things, all the cold which enters at the hall-door, falling to the floor, never goes up-stairs, but mixes and rises gradually with the air in the hall; and of course reduces the temperature of the lower part of the house. It is for these reasons, that whilst our dwellings are too warm up-stairs, they are generally too cold below, where warmth is most needed. Where there is no open stairway this is never the case. In order to remedy this evil, as far as practicable, you must hang a curtain all around the opening of the stairway, so as to hang down about two feet below the level of the under-side of the apertures E, Plates XII. and XIII., which let the air from the hall into the rooms. In this way, you see, you keep a full supply of the warmest air for the lower rooms; whilst none but that which is of a temperature two or three feet lower down, on a level with and below the bottom of the curtain, can get up-stairs.

In the course of my experiments and operations in ventilating old houses for several years, I have had many new brick chimneys built in them for the mere purpose of exhausting the air from rooms which contain no chimneys; but I find this now unnecessary. A *wooden* chimney, or air-shaft, is really better than a brick one for this purpose; so that if you have one *brick* chimney, in such a position that you can put your smoke-pipe into *it*, all the other exhausting shafts may be made of wood. Plank, or boards well seasoned and matched, and put together with thick paint, will answer the purpose admirably. These may be put up with much less trouble and expense than brick flues. They may be made to resemble a brick chimney, both inside and outside of the building, if that be desirable. A regular mantle-piece, chimney-board and valve, the same as seen in Plate II., may be made; and the chimney-top, which on any chimney must always be carried up as high as possible — say eight or ten feet above the peak of the roof — as well as the Emmerson cap, Plates XXVI. and XXX. For chimneys which have fire in them, those caps must be made of sheet-iron; but for wooden chimneys they may be of wood.

I have also, very lately, made another improvement in fresh-air ducts; and this is equally applicable to both old and new buildings. It is this: In a new house, before the hall-floor is laid, make the duct to run under the hall, directly across the building, open at both ends, which ends come through both foundation-walls, so that air will circulate freely *through* it; of course it must be placed directly under where the air-warmer is to stand; and there, if made of wood, hung to the joists. In framing the joists for the hall-floor, in a new house, considerable labor may be saved by *trimming* them in the exact place, and to the exact size required, before the floor is laid.

Then hang within this duct two horizontal swinging-valves, A, Plate XXXI., or Y, Plate XLIV., made of the lightest possible material; so that the least flow of air through the duct will open or close them. These valves are to be hung, one on each side of the opening which lets the air up to the air-warmer; and they are both to be so hung as to open *inward;* that is, toward the air-warmer. Then let the air come into the duct from either end, the valve nearest that end will *open*, and the opposite valve will *close*, and prevent the air from going out of it, and thus force it up the aperture through which the air-warmer is supplied. In order to facilitate the repairing of these valves, should they at any time get out of order, the best way will be to hang them at or near the ends of the duct, as seen in Plate XXXI.; or if this be impracticable, you may cut out the side of the duct at any point between the air-warmer and end of the duct, and place them there.

Having made these preparations, you now place the air-warmer over the air-aperture, (more particular directions will accompany the machine,) as also seen at Plates XXXI. and XXXII. Both the air-warmers have two smoke-pipes each. These are five inches in diameter, and should both be run up into a nine-inch pipe, (see the Plates XII. and XIII.;) and from this nine-inch pipe a common seven inch smoke-pipe will be run up-stairs through a smoke-pipe register (see Plate XXXIII., Fig. 2) into a dumb-stove, D, Plates XII. and XIII.; and from this dumb-stove into your chimney-flue. A dumb-stove is nearly as essential an adjunct to the warming of the building as the air-warmer itself. Indeed, I consider it of such importance as an economizer of fuel, that I never undertake the warming of a one and a half or two-story dwelling, nor allow my air-warmer to be used unless the proprietor agrees to have one put up; and I advise you to insist upon the same thing. The heat from this dumb or any other stove may be used and availed of with impunity so long as you have the copious flow of fresh air which one of these air-warmers will supply.

Now, pull open the fresh-air regulator, or register, lying at the bottom of the pedestal of the air-warmer, Fig. 3, Plate LII.; open the fan-registers, F, Fig. 3, Plate III., and pull up the sliding-valves in the chimney-boards, G, Plate II.; put fire into the air-warmer, and though the thermometer outside may stand below zero, your house will be warmed in half an hour. It is always best to have (besides this register or fresh-air regulator under the air-warmer) a good air-tight sliding valve, by which every particle of air may during the coldest nights be shut off, as the iron registers are not always air-tight.

You will have learned before this that I do not hold to the principle of distributing the warmed air throughout the several apartments of a house, by means of "pipes" leading from a "hot-air chamber." This is a most vicious as well as a dangerous practice. Hot air in contact with wood, even though separated from it by metal — even tin — will set a house on fire; and I wonder that the practice has been so long submitted to by owners and insurance companies. *The exhaustion principle* of distribution is the only safe one; it is far more easy, less expensive, as well as more efficient.

If, therefore, you are called upon to ventilate a building already erected, whether old or new, and the owner insists upon your making use of a "furnace" placed in the basement or cellar, lay a good stone or brick foundation, and set your furnace on legs or brick pillars, so as to leave a clear space of six inches *under* it, so that when the fresh-air duct, or ducts, bring the air through the basement-wall, it is left free to pass *entirely under* the furnace to the opposite side. If the furnace be any of the ordinary kind, of which we have a few left about the country, your fresh-air duct or ducts (if you have more than one) should, in the aggregate, bring within the chamber six or eight feet. Make your inclosure or chamber of brick and tin, and large enough to leave a six-inch clear space between it and the furnace, all around. If made of brick the wall should be not less than eight inches thick. Then, on the hall-floor, directly over the furnace, mark out the size which will let the six or eight feet, or whatever quantity of air you bring in, through; then mark out and cut an aperture the width of a brick (say four inches) *outside* of your first mark, and cut down through floor, joists and all. Then bring up your wall, so that four inches or the width of a brick will be between the inside of your floor-aperture and the inside of the wall; place your register, then, upon the top of the brick, so that its top will be flush with the top of the floor. You thus bring your whole quantity of air at once into the hall, making the whole house the "hot-air chamber."

If now you put in your fan-registers, F, Fig. 3, Plate III., and open your chimneys or flues — if they are not already open — it is easy to see that, if you have fire in your furnace, how the old and cold air will rush up these flues, its place being supplied through the fan-registers, until the apartment is filled with the fresh and warmed air; the old air having been chased out by the pure.

Where I have found two adjoining rooms, only one of which had a good flue in it, I have made a good job of warming the other by making an aperture of one hundred and forty-four inches close down to the floor through the partition; but then there ought to be fan-registers to let the warmed air from the hall flow into *both* rooms; if this be impracticable, one at the top of the *partition*-wall will be the next best thing to be done — especially if you can double the size of the fan-register in that room which you can connect with the hall. So, also, supposing these two rooms are distant from each other, put a register in the floors of both, and connect the registers by a good tin or tight wooden duct, *under the floor*, and the operation will be the same.

It does not matter much what *shape* your fan-registers may be; only let them be large enough and placed as high up as the ceiling or cornice will admit of. The leaving a door open between a bed and sitting or other room, in which there is a good exhausting-flue, will be ordinarily quite sufficient.

It can not be expected that I can anticipate all the difficult cases with which you may meet; but a thorough understanding of the physical properties of air, as elsewhere explained, patience and a liberal proprietor, will enable you to surmount all difficulties — *never forgetting that if you want any different air than what is already in an apartment, you must, the first thing, make room for it* by letting the same quantity out.

PLATES XIV., XV., AND XVI.

VENTILATION OF SCHOOL-HOUSES.

A, foul-air shafts or chimneys. B, fresh-air ducts. C, air-warmers. D, open base. E, entrance of foul air into shaft. F, joists. G, space between joists and ground. H, sliding-valve to regulate quantity of foul air to be let up shaft. I, box-stove to be set between recitation-rooms if necessary.

I give on Plates XIV., XV., and XVI. plans and sections of two school-houses. These are mere random designs, which I supposed would probably be as convenient for common-school houses as any others. So to those who understand what I have already written, they explain themselves.

You will perceive that the ventilation is carried on in the same way as a dwelling-house without any basement. The air enters the fresh-air duct at B, and flows under the air-warmer, C; it fills the room and drops under the floor all around the apartment through the open base, D, and is thence drawn through the space, G, between the joists, F, and the ground, up the foul-air shaft, A. This flow of warmed air under the floor effectually cuts off the cold from the under side, and, of course, the feet of the scholars. The floor is diminished about two inches all around the room, as in the dwelling-houses, and as seen on Plates I., II., III., and IV.; the open or perforated base is not necessarily made of iron; a perforated *wooden* base is just as effectual, only be sure to have a sufficient number of holes or apertures to allow double the quantity of air to go through the base that you bring in through the fresh-air duct, B. These perforations must be begun on that end of the school-room which is *opposite* to the shaft, and continued down equally on both sides of the apartment, until the aggregate or sum total of the quantity above mentioned be obtained. These perforations may most easily be made with an inch-auger or bit, and may be so disposed as to be ornamental.

The ceiling should on no account be higher above the floor than ten feet — nine or even eight feet would be better — for not only is a high room much more difficult to *warm*, but it is also, with every foot in height, more difficult to *hear* in. This is not all. The smaller a room is, the more rapid and effectual will be the ventilation.

The impression has always been, that school-rooms, of all others, should be *high*. This is true when there is no ventilation, but when there is ventilation, the very reverse is the case. Suppose we take two vessels — say, a whole puncheon and a half-puncheon — it is evident that, with the same sized stream of water, the half-puncheon will be filled twice to that of the whole puncheon once. This is exactly the operation which takes place in the ventilation of two rooms, the one being eighteen feet high and the other nine feet; the low room gets, in the same period of time, double the quantity of fresh air that the other does. The low room, you will observe, will hold the same number of people as the high one, and here you have the extraordinary advantages of *hearing* better, of having double the quantity of fresh air, and of being warmed in a winter's day with less than half the expense of fuel!

Plate XVI., Fig. 1, also explains itself, showing that the fresh air may be brought in under one of the windows, if desirable. Fig. 2 also shows how the quantity of foul air going up the shaft is to be regulated.

The air-warmers which I have devised for school-houses may be set in any upper story with equal facility as in the lower story; so that there is no more need of these expensive furnaces and a fireman — the teacher and children supplying the fuel as required. The room is warmed in less than half the time in the morning, and there need be no more fuel consumed than will last till the time of closing.

These plans, I believe, are intended to be drawn on a scale of eight feet to an inch. The buildings, as will be seen, are only one story high, and have no basements.

It is utterly impossible to make children comfortable in a Canadian climate, in a room where the ceiling is very high, and it will be allowed on all hands that a child can not learn his lesson when he is shivering with cold, and he will always shiver if *his feet* be cold. The first thing, therefore, to do, after giving him a full supply of fresh air, is to secure the scholars against cold feet.

The effect of exhausting the out-going warmed air under the floor will be not only to cut the cold off from the feet completely, but will impart to the floor whatever warmth may be remaining in the air.

The prevalent idea is, that the floor of a room can not be cold if it be a good tight one, as well as a tight foundation under the building. This is altogether a mistake. Put a thermometer under the floor of a building which has no foundation-wall at all, and under another with the best air-tight foundation, and you will find that both will mark precisely the same. The one space under the floor is just as cold as the other. I have, many a time, (and it is the experience of every Canadian of observation,) seen water, accidentally dropped upon the floor of a stove-heated apartment, instantly congealed, and this when, at the ceiling of the same room, the thermometer would mark 100°. The air-tight foundation, however, gives you this advantage, (and one of the utmost importance over a building without any foundation at all,) that you are enabled, by the system of ventilation, to draw all the *cold from under the floor* and substitute warmed air, so that if the outside air be at zero, there will be a difference in the temperature of 40° or 50° between the temperature under the floor and that outside the building, and frequently much more. You can imagine, therefore, what an advantage the system of warming by the ventilating process possesses over any other.

Now as regards the ventilation or change of air. I have made the ceiling of my school-house nine feet high. This room, then, 40 x 20 feet on the ground and nine feet high, will contain seven thousand two hundred cubic feet of air. The air-warmer which I have devised for a room of this size (and even a larger room) is quite small and portable, taking up room on the floor only thirty by eighteen inches. This machine will throw into a properly exhausted room, on an average, during over five winter months, five cubic feet of air warmed up to 100° per second, equal to three hundred per minute, which will change the air in the building every twenty-four minutes. In other words, there will be no air in the building half an hour old.

So far, then, as the mere VENTILATION goes, it will be admitted that this is quite sufficient.

Now as to the WARMING.

In order to judge intelligently upon this part of the subject in hand, it will be best to inquire into the three different ways of warming our schools, now extant.

The common stove. The steam or hot water. The furnace.

I admit, and must now be allowed to repeat, that any *active* heat to the feet, and especially to the feet of children who are so frequently running out into the snow, is injurious. All that is required to keep us warm is *to keep the cold away from our feet;* in this lies the great secret of keeping the human body warm in a cold climate. If our feet be cold, we are cold all over; if our feet be warm, we are warm all over, let the temperature of the air in which we sit be what it may. I pronounce it an impossibility to be comfortably warm when we have cold feet, even supposing we sit in air up to 200°.

We have already seen that the floor of a stove-heated apartment is frequently so cold as to congeal water, and must therefore have the same effect upon the soles of our boots and shoes in contact with it as if we were treading upon an inch and a half thick sheet of ice.

By many experiments I have ascertained that on a zero day, in a school-room sixteen feet in height, having two sides exposed to the outside air, built with brick and plastered upon the wall, the mercury will fall $4\frac{1}{2}°$ in every foot from top to bottom, so that if the thermometer stands at 100°, which is the usual temperature at the ceiling of a stove-heated room: in such weather and in such a room the mercury will stand not higher on the floor than about the freezing-point. [In a dwelling-house where the heat above is kept up for the whole twenty-four hours, and where there is constant or even occasional communication with the basement, it is somewhat different.] Nothing can, under such circumstances, warm the feet, and consequently the person, but the keeping the cold from the *bottom* of the floor-boards, and yet the teacher and others, in the vain attempt to warm the inmates, "fire-up" to such a degree as to keep the *heads* of the children in an atmosphere of 80°, not considering that their *feet* are all the time upon an ice-floor.

With respect to steam or hot-water heating, this is only worse than stove-heating, inasmuch as this mode of heating does not admit of the removal of one particle of air from the room, which the stove-heating does, to the extent of the combustion air at any rate.

The furnace. Air can not be heated to over 140° without injury, and, in order to *force* up air into an apartment, by forcing out of the cracks and crevices of the room, and especially the ceiling or upper part, a sufficient quantity to allow that of the furnace to enter the room, the air is brought in through the registers at a temperature of from 200° to 500°, and sometimes much more. The destructive effects of such air upon the health of children is beyond description. The first symptoms are a dryness and soreness of the throat, inflamed eyes, and cough from the elimination of sulphur from the overheated metal and brick-hot air-chamber—prostration of intellect and sleepiness.

But it may be asked, "Why, if good chimneys and exhausting flues be erected with these modes of heating, the operation will not be as good and effectual as yours?" We all know that these provisions have not hitherto been made, but I answer, suppose they are hereafter always made?

In this climate we can not afford the fuel to exhaust the room at the *top*, and if we could, the disparity of temperature between the head and feet of the inmates would be increased to such an extent that the cold at our feet would be insufferable. "But suppose the exhaustion be made on a level with the floor, how then?" I answer, that as regards that branch of the subject which we are now discussing, "the warming of a school-room," the pupils would be no better off than with either of the other modes of warming.

Air at 90° in an inert state will feel insufferably hot, but put that same air in swift motion, and it will *feel* cold; how much worse when, instead of 90°, the air is down to 40°? I have shown that under certain circumstances, when the air is at 100° at the ceiling of an eighteen feet high room, it will be about 32° at the floor; but suppose we, for argument's sake, allow it to be up to 40° or even 50° at the floor, set this air in as rapid motion over the top of the floor as a good three-feet flue would do, and the extraction of heat from the feet (the blood being at 98° or 100°) will be more rapid than if the bottoms were immersed in inert air upon a floor as cold as ice; whilst the head would be in a body of inert air at say 70°, the feet would be in a cold bath. I put it down as a postulate that no man can feel comfortably warm during our winters in any room *over* the floor of which there is a local *current* of air.

Now, then, contrast with these several modes, warming by the ventilating process.

The cold air, being drawn out of the space or inclosure below the floor by the shaft or chimney, and the vacancy thus made supplied through the perforated base by the warmed air of the room above, as already described, the whole space below, instead of being down to zero as before, will come up to a temperature of 40° or 50° at least above that point, and the floor-boards, being between these two bodies of air, must be considerably above a freezing point — the whole space, indeed, below the floor will become almost warm enough to live comfortably in, except for the local current. But the main benefit, so far as the warmth to the feet is concerned, is derived from the local current of air flowing to the shaft being transferred from the *top* of the floor to the *under side*.

The motion of air in a room warmed how you like is this — it goes from the air-warmer directly upward, and thence spreads over the *whole* ceiling, thence down the side-walls *down* to the floor; from this point in a stove-heated room, if there be no provision made for ventilation, the air flows from all sides of the room to the stove or stoves, sweeps up the hot metal, and again takes the uppermost place against the ceiling, and so on as before. In a hot-air-heated room, after falling down the side-walls, it converges to the hot-air registers, and is carried to the ceiling with the stream of hot air, and so on. In both cases there is a constant current of air close over the top of the floor, which, having now lost much of its temperature, tends to cool the feet. In both these cases, if provision is made at the level of the floor for the exhaustion of the air, (which being necessarily at one end or side of the apartment,) its rapid flow converging from all points to the flue, increasing both in quantity and velocity and losing its temperature, renders it, in very cold weather, impossible to sit at the flue-end of the room without suffering from cold feet.

Now let us see the operation under the ventilating system of warming. The air from the air-warmer ascends to the ceiling, and flows down the several side-walls to the level of the floor, as in the other cases; but here, instead of flowing *over the floor*, descends, in consequence of the opening of the shaft being *under* the floor, through the open base, and flows to the ventilating flue *under* it—leaving thus all the air lying *upon* the floor in a perfectly inert state. I need say not another word as to the superiority of the one mode of guarding the feet against cold over the other.

It is usually estimated that what with the destruction by the lungs and cutaneous transpiration and other emanations from the body, every child will contaminate, so as to render it unfit for healthy breathing, seven cubic feet per minute. A school-room 40 x 20 feet on the ground and eighteen feet high, contains fourteen thousand four hundred cubic feet, so that in about twenty minutes one hundred pupils will contaminate this whole body. The opening of a door (or two doors, if they are both on one side or one end) of a room destitute of any open flue, will cause no *change* of the air in such room to speak of, because if the air can not go out, it can not come in a room, but we will allow, from this cause and from the air which comes in through the cracks and crevices, one cubic foot of air to come into such a room every minute. This fourteen hundred and forty feet every twenty-four hours will renew the air only once in ten days, so that for all practical purposes the ventilation from doors and cracks and crevices amounts to nothing.

If, then, the whole body of air within this school-room be contaminated in twenty minutes, what state must the air be in at the end of even one day of eight hours when it must of necessity have been subjected to the same deterioration twenty-four times! What at the end of one week, (the same body of air having been carefully locked up every night for use the next day,) when the emanations from the lungs and bodies of these hundred children will have been added 144 times! In four weeks, 576 times, in a twenty-weeks winter, 11,520 times! The very walls, and especially the ceiling of such a room become so impregnated as to affect visitors' olfactories to such an extent as to affect the stomach. This fact is within the experience of all persons who are in the habit of occasionally entering an old school-house, even when no pupils are within it. Such appears to be the subtilty of this poison absorbed by the walls, ceiling, and floor of a school-room, that a whole summer's sweep of air through it has no perceptible effect in its extraction; the very day that it is again closed up for the winter, the same smell of corruption is just as apparent as the year before.

"The condensed air of a crowded room," says one of our first chemists, "gives a deposit, which, if allowed to remain a few days, forms a solid, thick, and glutinous mass, having a strong odor of animal matter. If examined by a microscope, it is seen to undergo a remarkable change. First of all, it is changed into a vegetable growth, and this is followed by the production of multitudes of animalculæ; a decisive proof that it must contain organic matter, otherwise it could not nourish organic beings."

Dr. Hiller, Secretary of the Metropolitan Medical Association, says: "In consequence of the ill-construction and bad ventilation of the school-houses in and about London, seven thousand children, between the ages of five and fifteen years, annually lose their lives, from these causes alone."

Hear, again, another veteran in the medical profession:

"The occupation of such rooms being the lot of the larger portion of the rising generation, who can wonder that our race is degenerating in physical powers? Who can doubt that such a state of things prepares the soil and sows the seeds which in due time spring up into that luxuriant harvest of ailments and complaints which is reaped by the victims of our school-rooms?

"The stupefying effects of dark venous blood poured through the brain is unhappily most apparent where there is expected to be the highest degree of mental activity. School-rooms are never provided with due means of ventilation, by which a constant supply of pure air may be maintained; and the inattention, dullness, and sleepiness of pupils are but the natural and inevitable consequences of taking into the system a vitiated and poisonous atmosphere. It would be wise for teachers who are afflicted with pupils of dull and stupid intellect, to inquire how far the stimulus of pure air might be advantageously substituted for flogging.

"The tender, sensitive child, that sits and reads, and learns his lesson, and perhaps can not learn his lesson, and stupefies and pines, and droops, and may be has scarce a smile to expect when his task is done, yields, day by day, to his atmospheric foes. Day by day, and as he loses the first start of life, his lungs play less freely, his blood circulates more slowly, his chest contracts, his limbs pine away, his digestion is disordered, and before long he is delivered to the tender care of the man who gallops in every other day, sends whole bales of pills and draughts, and soon settles either the life or the constitution of his unfortunate patient.

"It is needless to urge that danger to the health and life of the child is so remote and trifling as to be unworthy of consideration. The reverse is the case. Instances are constantly occurring in which the seeds of disease are gathered in the close and polluted air of the school-room, to ripen into premature decay and an early death. Many parents can call to mind the frequent complaints of their children, who have returned from school-rooms, feverish and pale, laboring under a depression of spirits and lassitude of body. A passing emotion of compassion may have attributed their appearance to confinement and study, neither of which is productive of evil effects, unless accompanied by an atmosphere rank with impurity, habits opposed to cleanliness and health, a loss of comfort and necessary recreation.

"In a school-room with no means of ventilation, and containing from fifty to one hundred scholars, the air inhaled by each different pair of lungs loses its vital properties, and becomes loaded with the impurities and infections thrown off from numerous systems. To contend that there is in this no danger to the health of the child is folly. The temporary symptoms of suffering may disappear with the habit which occasioned them, but the tendencies of disease linger in the system, awaiting some predisposing cause to develop their active strength, and hurry their victim to an untimely grave.

"These statements are no exaggeration of the evil, for exaggeration is impossible. Still, the evil is allowed to exist, because its first manifestations are not in a form that appalls and terrifies. Its approach is slow and insidious; the operation proceeds in secret. At length the frame, racked with pain, a mind debilitated, unbalanced or diseased, powers of usefulness or enjoyment destroyed, are the fatal results of a few years spent in a crowded and heated school-room. For all these consequences the prevention is of the simplest character. The most ordinary mechanical contrivance will insure pure air to the child and happiness to the man. That is a costly economy which sacrifices sound health, and disregards the danger of disease, to save a trifling expense."

PLATES XVII., XVIII., XIX., XX., XXI., XXII., XXIII., XXIV., XXV. (Scale, 12 feet to 1 inch.)

VENTILATION OF PUBLIC BUILDINGS.—JAIL.

A, foul-air shaft. B, fresh-air ducts. C, air-warmers. (C, on second and third floors represents registers through which the smoke-pipe or pipes, as well as the warmed air, are to come from below.) D, cells. E, reverse-flues. F, doors of cells. G, foul-air chamber. H, apertures over cell-doors, to admit warmed air, Plate XX. I, gallery, Plates XX. and XXV. K, stove in foul-air shaft, to rarefy the air, Plate XXI. L, aperture for entrance of foul air into shaft, Plate XXI. M, top of foul-air shaft, Plate XX. N, brick foundation upon which air-warmer stands, Plate XX.

I regret that imperfections and omissions in some of these Plates, render it necessary for me to supply by writing what would have been much more intelligible to the mechanic in drawings.

The first omission which I notice is that of a drawf-wall across under the corridor, immediately back or on the left-hand side of the foundation of the air-warmer or furnace, N, Plate XX., and C, Plate XVII.; so as to prevent the fresh and cold air from B passing the air-warmer and mixing with the foul air at G, Plate XX. and Plate XVII. You will have seen, in the discussion of this subject, that *no air other than that which has come through the air-warmer, must be allowed to go up the foul-air shaft, or the whole operation will be spoiled.* Hence the absolute necessity of an air-tight wall at the back end of the air-warmer, quite across the whole width of the space under the corridor.

Whatever air-warming machine may be used, it must be sustained by legs or pillars, to the same *height* as you see N, Plate XX.; so as to allow of the fire being supplied with fuel *above* the level of the floor of the corridor.

Instead of bringing the fresh air in at B, as you see in Plates XVII. and XX., I recommend the following improvement:

Before the foundation-walls of the building be laid, a good, substantial "plenum fresh-air duct," Plate XXXII., say six feet wide and about two and a half feet, or as deep as will carry sixteen feet of fresh air, must be made across the whole building from G to H, and of course directly under both the air-warmers, C. This duct must be laid so low down in the ground as not to interrupt the free flow of the foul air *over the top of it*, through the arches, G, Plate XXIII., upon which the two blocks of cells are built. As the air from B, Plate XVII., and B, Plate XX., must of course intersect the air from both ends of the large cross-duct laid from G to H; and as the side of this last would have to be opened under or nearly under each air-warmer, so as to allow of this intersection; and as we require these two (both leading from the same point of compass) to carry only eight feet of air each, I recommend that they should be built from B, to their junction with the large duct — only about 4 x 2 feet. The whole thing would then represent two wings, F, Plate XXXI., both intersecting the large duct, one under each corridor. It is not important that these wings should join the large cross-duct *exactly under* the air-warmer; yet it may, perhaps, be as convenient to allow of their junction at those points as at any other. You will, perhaps, perceive that I have all the while imagined this building to stand north and south; the foul-air shaft, A, south. The plans and other Plates connected with them are drawn upon a scale twelve feet to an inch.

Having made these preliminary observations in explanation, we will now proceed:

As with private so with public buildings; the diversity is so great, no two being alike, I can do no more than select one kind in order to illustrate their ventilation. For this purpose I select a jail.

The security of prisons being the main object, it must be manifest to every one that the placing the cells in a block, in the centre of the building, is better than to place them against the outside-walls. I have imagined a block of thirty-six cells, built back to back, three stories high — twelve cells in each story. There should be a ten-foot corridor all around the whole block, (through a misunderstanding of my draughtsman, I perceive the drawings represent it only on two sides and one end.) By the Plates, there is a basement of three or four feet (and it may be deeper if necessary) under the whole of this building; this basement (with the exception of the space necessary for the brick or stone arches to bring in the fresh air as already mentioned) is the foul-air-gathering chamber for the whole building, and serves the same purpose as the spaces under the floors in the description of the ventilation of a dwelling-house, Plate X.; and as the foul-air-gathering ducts, more particularly described on Plates II., IV., V., VI. The fresh air is brought under the machines, which warm it, by the large duct laid under the building, from G to H, Plate XVII., as already described (assisted by the two smaller ones, F, Plate XXXII.) from B, Plates XVII. and XX., as indicated by the arrows.

The whole block of cells, you will perceive, is built upon (or rather across) two arches, G, Plate XXIII., each of which should be of a capacity sufficient to carry at least ten feet of air. It is into these arches that the whole of the foul air from both of the half blocks of cells above them respectively must be drawn; through the "reverse flues," the cross-sections of which are marked E, upon the plans of each story; and the longitudinal sections of which are E, on Plates XXI. and XXIII.

Let us now follow the air from its entrance into the fresh-air ducts, as before described, until it goes out of the top of the shaft, M, Plates XX., XXI., XXII., and XXIV. Each air warmer or heater must be supplied with eight feet of air, and it must be of sufficient power to bring that quantity up to the temperature of not less than 120° of Fahrenheit, when it is flowing at the rate of five feet per second, and the thermometer outside is standing at zero. The air is now forced and drawn up through and around the air-warmer, C, and fills the corridor from top to bottom—flows through the apertures over the cell-doors, D, at H, Plate XX., fills the cells, enters the top of the reverse-flues, E, which open near the floors of the cells, flows down them (as indicated by the arrows) and into the arches, G, Plate XXIII., fills the whole space of the basement, including those under the day-rooms, (the floor being cut away in the drawing to show the arrows, which indicate the course of the foul air,) toward and up the shaft at L, Plates XXI. and XXIV., and of course out of the top of the shaft.

The "reverse-flues" should contain each not less than one hundred inches, and they are best when made of inch boards, merely nailed at the sides, and set up perpendicularly, one upon the top of the other as the walls progress, and working the brick, forming the division walls of the cells, around them. This is the cheapest as well as the best mode, (these boxes are left in the wall, of course;) you thus save a great deal of trouble in keeping the inside of the flue to a face in plastering the inside, and security against its being filled up with mortar and brickbats. The division-walls between cells should never be less, if built of stone, than twenty inches thick; if of brick, twenty-four.

Let us now make a little calculation. We will suppose eight feet of air to flow through one of the air-warmers at the rate of five feet per second. That, when it comes out of the air-warmer, it has by expansion increased to ten feet. This ten feet, multiplied by the velocity of five feet per second, is equal to fifty feet per second, which, multiplied by the number of seconds in a minute, (sixty,) is equal to three thousand cubic feet of warmed air per minute. Each air-warmer has to supply the half-blocks of eighteen cells, so that the three thousand feet, being divided by eighteen, gives nearly two hundred cubic feet to each cell per minute. This is nearly double the quantity that can apparently pass down the reverse-flues; but inasmuch as the velocity of air always increases, the nearer it approaches the chimney or air-shaft, where there is little or no friction, as in this case, we may safely calculate that one hundred and fifty cubic feet per minute will pass down each reverse-flue; this one hundred and fifty, multiplied by the number of cells in the half-block, (eighteen,) is equal to two thousand seven hundred cubic feet. The remaining nine hundred feet per minute I leave for the diminution of volume by a decrease of temperature, for diminishing the quantity of fresh air coming to this air-warmer in cases when the weather is so cold that the air-warmer can not sufficiently warm the full quantity, and also for keeping a surcharge or pressure, if I may so speak, within the whole building. There must be a *sliding*-valve (besides the self-acting valves, A, Plate XXXII.) constructed at the entrance of the fresh air, (inside the wall for convenience,) by which the quantity of air admitted to the air-warmer may be regulated and entirely shut off at night, if ever found necessary.

The rarefying stove, K, Plates XXI. and XXIV., is to be placed near the top of the shaft for the purpose of rarefying and thus quickening the draft of air up the foul-air shaft in hot, calm, and sultry weather in summer; and the flue of the foul-air shaft must of course be such in size and shape as to admit *past*, on both sides of this stove, sixteen feet of air, eight feet on each side. The drawing represents the flue as 4 x 5 = 20 feet, but the arch upon which the rarefying stove is to stand, say twelve inches wide, must of necessity be thrown entirely across the shaft; it will be more economical of space to make the flue somewhat longer one way than the other. The stove, as shown in the drawing, is too large — it need not be more than twelve or fourteen inches wide.

In the three jails which I have ventilated upon this principle, I have, of course, used my own air-warmer, as the ordinary furnaces would take up more than double the room, consume more than double the fuel, cost double the sum, and will not warm more than half the quantity of air. My basement-heater or air-warmer is supported entirely *above* the fresh-air duct, so as that the fire may be fed from the floor of the corridor, and is covered with tin, (though a twelve-inch brick wall might be as well.) There are four smoke-pipes of five-inch diameter, each run up through register to near the top of the corridor, and then join a twelve-inch horizontal pipe which runs into the shaft, as seen in Plate XX.

If I have succeeded in making myself thus far understood in the ventilation of a jail, it can not be a difficult matter to apply this operation to any public building.

It is an easy matter to make a sub-basement, especially as it should be as nearly above ground as possible. It need not be more than from two to four feet deep. Then connect your foul-air shaft or shafts with this sub-basement, the "reverse-flues" also connecting this apartment, and exhausting the air from the cells into it, similar to these we have been considering in a house. Construct as few foul-air shafts as possible, and get as much fire into those you have as you can.

The exhausting-shaft of a jail must not be less than sixty feet high; seventy or eighty would be better.

It is easy to perceive that you may have as much water-closet room about these shafts as will be needed; only make access to them easy and convenient for the removal of the *sordes*. Use plenty of lime, plaster-of-Paris, or ashes, according to the number of inmates. Allow no water to be thrown down the pipes, except that which is naturally incident to the privy.

The large rooms on each side of the shaft, A, Plate XVII., are intended for day-rooms, and may be thus, or in any other way, divided in every story, if required. By extending the *side*-walls of the shaft up to the wall which divides the block of cells from the day-rooms, there may be a separate water-closet for the prisoners on each side of each block of cells. (See chapter on "Dry Closets.")

PLATES XXVI. AND XXX.

FOUL-AIR SHAFT OR CHIMNEY.

Plate XXVI., B B, cap or canopy of the Emmerson cap. C D, the course which the wind or moving air is deflected by the shape of the outer part of the brick-work.

Plate XXX, A, wall of shaft. B, inside of shaft. B B and E E, on Fig. 2, inside of sheet-iron flange, which forms, when put to its place, the inside of the flue. E, outer part of the sheet-iron flange. D D, cap or canopy of Emmerson cap.

This part of the building is the most important feature in the whole structure. It is that upon which all the rest depend. It may indeed be said to be the "wind-pipe" of the ventilating system, as that organ is of the human system. Without this organ no animal can breathe, and without this no house can be made to breathe. Seeing, then, that such is the case, it is all-important that the principle upon which it works should be thoroughly understood by the bricklayer or artisan whose business it is to construct it.

We have all learned and we know as a matter of fact, that the higher a chimney is the better it will "DRAW." Recollect, I do not say the *longer* a chimney is, for if you dig down into the ground forty feet and build up your chimney from the bottom, fifty feet long, the top being then ten feet above the surface of the ground, this chimney will "*draw*" no better than a ten-foot-high chimney commenced at the surface of the ground. We know also, as a matter of fact, that the higher the wind blows, the better a chimney will "draw." We know furthermore that a chimney situated on the top of a hill will "draw" better than one situated in a valley or amongst and under cover of taller buildings, or other impediments to the free circulation of air.

It follows, therefore, that the height from the surface of the earth, the force of the wind, and freedom from impediment in its circulation, must have something to do with the draft up a chimney or flue.

Air is so thin, so ethereal and volatile, that it is never at rest; the waving of a hand in a roomful will disturb every atom within the apartment; especially is it affected by heat. Place a coal of fire in a cold room, and that instant every particle is in motion. It is withal so compact that no one particle can be taken from or added to any body of it without disturbing the whole, like as if it were an ethereal jelly, if I may use the expression. Any attempt to draw away any part of it is followed by the whole body moving in a mass. The number of natural disturbing causes are so various and so infinite, that it is impossible to guard against them, and the only power we can ever possess over its movements is to direct them.

Heat, natural and artificial, is the most powerful of these disturbing causes; and we have learned that under its influence air will in general rise *upward*. It was only after this discovery that chimneys were built in dwellings. Air may of course be moved by mechanical means, but with this we have nothing to do at present.

Knowing, then, these two facts, that air is never at rest, and that by means of heat we can give it one direction at least, we are left these two facts at any rate upon which, by our mechanical contrivances, to insure a continuous draft up our chimney or foul-air flues, and as a consequence, in a horizontal or any other direction which by our mechanical constructions may be determined upon.

I have said, and indeed it is the fact, that the higher a chimney is carried above the general surface of the ground, the better it will "draw." The reasons of this will, upon reflection, readily appear: First, The motion of the air is retarded by the inequalities and friction of the surface of the ground or water over which the lower strata or the bottom part moves, whilst the upper strata or top part moves on uninterruptedly, giving you a smart, brisk wind, the farther you ascend, (every sailor will tell you that the first sail which receives the breeze after a calm is that one which is the highest,) so that, whilst the top of a tall chimney is swept by a rapid and direct current of air, the short chimney has to encounter not only a much slower movement, but also a much more irregular one, the body of air here being broken up by friction and irregularities on the surface of the ground, which by necessity causing a *rolling* motion and a tendency downward and into the top of the flue, acts still further as an impediment.

Secondly. A heavy wind over the top of a chimney has the effect of taking off it a great part of the weight of the superincumbent body of air which in a calmer day presses with the same weight upon the column of air within the flue of the chimney that it does upon the same area of surface any where else.

Thirdly. The tendency to a vacuum on the *lee* side of the chimney is greater in proportion to the velocity with which the wind goes, and this vacuum near the top is more readily filled *from* the inside of the flue than from the air outside which has just shot past and is already at a much greater distance from this vacuum.

Having now given reasons why a tall chimney will draw better than a short one, and whether these reasons shall be considered conclusive or not, we all know it to be an incontrovertible fact, and seeing that every thing depends upon its draft in the ventilation of buildings, it behooves us to insist upon carrying out above the roof our chimney-tops to the utmost possible height.

But in order to insure success in the full and rapid exhaustion of our building, there are other considerations only second in importance to the height, which are to be noted in their construction.

The top of the flue must be *smaller* than it is in the main body or trunk, (Plate XXVI.) The reason of this you will better comprehend by reverting to the illustrations which I have elsewhere given of the movement of air, by its comparison to water.

Go to a running stream of water and throw three chips upon the surface, one on each side of the stream near the edge, and one in the middle. The middle one will outrun those near the sides. There is no other cause for this than the friction caused by the gravel or other substance of which the sides and bottom are formed, impeding the progress of that portion of water which runs next to them; so also will it be if you sink any substance which is a little heavier than water, bulk for bulk, so that it will sink and float near the bottom of the stream, its motion will be retarded by the same cause—friction.

Let us apply this philosophy to air going up a chimney-flue which is as large as, or larger at the top than the belly or main trunk. Air will always tend toward the centre of the draft if not turned away by any impediment, so that with such a chimney-flue the outside of the body of air will rub hard against the brick or other material of which the inside of the flue may be formed, and the consequence of this will be a strong friction from the very bottom to the very top of the flue, and must cause the same impediment to the motion of the air that the sides and bottom of the stream do to the water.

Now if we *contract* the opening at the top of the flue one fourth or even one sixth, which will be about the proper proportion in general, it is obvious that the tendency of the draft will be *inwards* and *from* the inside wall of the flue, leaving a body of comparatively dead or inert air between the inside of the flue and the stream of air going up, the friction being thus almost entirely done away with, being merely air against air.

I never plaster the inside of my flues, but have the bricks carefully laid, faced, and "pointed" in the same way as the outside or faced side of the outer wall of the building. The inside of the flue thus made soon gets glazed (especially if there be any smoke in the shaft) as smooth as glass. More or less of the plastering on the inside of our plastered flues will drop off, and besides will endanger the building by taking with it part of the mortar which should remain between the bricks.

I make use of the "Emmerson cap," (C D B, Plate XXX.,) because I find it a valuable adjunct to the draft of a chimney, but am not quite clear that I fully comprehend the principle upon which it acts. I find that on a calm day it is of little use, and infer from thence that the slope from C to D is intended to give a direction to the air when in motion, *upward.* This must, of course, create a vacuum *over* the *top* of the shaft, and thus draw the air *directly* out of it. The air, by the slope from C to D, being once thrown up against the under side of the cap B, can not, of course, leave it until it passes the point E, (Plate XXVI., Fig. 1,) when the air will be entirely clear to mix with the outside body. The cap B, Plate XXVI., and D, Plate XXX., acts the important part of keeping the weight of the body of air above from interfering with the operation going on *below* it, and forms a conduit for the air which is impinging upon the under side thrown upward by means of the slope from C to D. The air from the flues of the shaft must therefore come up and fill that space which lies between the top of the shaft and the line of the bottom of the body of air thrown up against the cap B, Plate XXVI., Fig 1, and D, Plate XXX., Fig. 1, which here will probably be from D to some point between B and E, Plate XXVI.

Whether right or wrong in this theory, the *fact* of the value of this "Emmerson cap" is indisputable, and the inventor, the late Mr. Emmerson, of Boston, is deserving of the gratitude of all generations to come.

On Plate XXVI. I have endeavored to represent a cheap mode of making this cap by dispensing with metal except for the cap or canopy, B, and the iron rods for its support, the feet of which are worked into the brick-work, but the other, on Plate XXX., is that which I recommend as superior, inasmuch as this is a complete protection to the whole chimney-top, which, if painted now and then, will last, and preserve the top of the chimney as long as the building itself. I take it for granted that you understand that it is to be made of sheet-iron, and the canopy, D, strongly supported by at least four substantial iron rods—six or eight of these iron standards will be better. By looking at Fig. 2, Plate XXX., you will see that there must be an *inside* part, E E, B B, also to this cap, the top edge of which must, of course, be riveted to the top edge of the outside flange, C C. The inside part, E E, B B, must be not less than twelve inches deep, and of a size which will just crowd *into* the chimney-flue. For a chimney-shaft whose flue is 4 x 2 feet, as represented by the plate, this inside part, E E, B B, should run down into the flue eighteen or twenty-four inches, but for an ordinary dwelling or school-house, twelve inches will be sufficient.

PLATES XXVII., XXVIII., XXIX., AND XXXVI.

VENTILATION OF A CHURCH.

A, foul-air shaft. B, fresh-air entrance, Plate XXXVI. C, air-warmer. D D, in Plate XXXVI., air-warmers. E, wall across church to form foul-air chamber. F, foul-air chamber. G, joists. H, firring. I, truss-beam or girder to sustain joists. K, pillars in basement to support girders. L, plastering of basement-ceiling. M, pews. N, raised floor under the blocks of pews, (four inches.) O, open base. P, basement-floor. Q, top covering of fresh-air duct. R, side of fresh-air duct. S, entrance of foul air into foul-air shaft. T, aisles. U, chancel-floor.

The three Plates, XXVII., XXVIII., and XXIX., are the plan, cross-section, and front elevation of a church.

Plate XXXVI. is a perspective view of the principal floor and basement, drawn upon a very much enlarged scale, in order to show more clearly their several parts as constructed for ventilation. These are so plainly shown, it appears to me, that I can add little to its elucidation, further than to direct your attention to the "references."

The mode of building churches with basements under them has, of late years, become so common that I have endeavored to give such a view as will enable you to comprehend how the building may be ventilated with that appendage under it. You will observe that the tops of the joists, G, Plate XXXVI., are not let down flush with the top of the truss-beam or girders, by four inches, and also that there is firring, H, of four inches in height laid *across* the joists, raising the floor of the pews eight inches above the top of the truss-beams or girders, whilst the floors of the aisles, which are laid upon the joists themselves, are only four inches above the truss. The floors of the blocks of pews being thus four inches above the floor of the aisles, you are enabled to get the out-going air easily under the blocks of pews, through a perforated base, O, which may be made either of iron or wood.

I will suppose the nave of this church to be 60 x 40 feet. Then 40 x 12 inches in a foot makes four hundred and eighty inches across the church.

This again multiplied by four inches, the height of the bottom of the firring above the top of the truss, will give nineteen hundred and twenty inches, or above thirteen feet of air, which can flow over the truss-beam.

Supposing the ceiling of this church to be perfectly flat and twenty feet above the floor, (and it should not be higher, notwithstanding the drawing,) it will contain forty-eight thousand cubic feet. The two No. 2 air-warmers being supplied with seven feet of fresh air, (three and a half feet each,) which, flowing at the rate of five feet per second, will give thirty-five feet per second, which again multiplied by sixty, the number of seconds in a minute, will throw into the building twenty-one hundred feet of air per minute. The forty-eight thousand feet now divided by the twenty-one hundred, the church (supposing the foul-air shafts to do their duty) will be filled and emptied in less than half an hour.

The flues in the foul-air shafts should be equal to carrying out fifty per cent more air than you bring in. If, therefore, you bring in an aggregate of eight feet of fresh air, your foul-air shafts should carry out twelve feet, in order to allow for the expansion of the air and the *additional* air which will always come in through cracks and crevices. These shafts, standing upward of forty feet high from their foundations, will easily accomplish this work.

You will see by the plate that I have supposed the building to stand north and south, and that the fresh air for the air-warmers should be supplied by only one duct, and that from the west; but having discovered the superiority of the plenum fresh-air duct since the artist made the plate, I now recommend you to use either the wooden one, Plate XXXI., or the brick one, Plate XXXII. The whole three wings, F, G, and H, to carry eight feet each. If you are obliged to make use of a furnace or furnaces to warm the air, the place is within the foul-air chamber, F. A passage may be made into it either through the wall, E, or from the rear end of the church, and a room of sufficient size be made in this foul-air chamber—only be sure to make all air-tight, so that not a particle of the foul air can come into it, nor a particle of the external air can mix with the foul air. The flues of the shafts, A, will be about 3 x 2 feet, or say two and a half feet square.

This wall, E, is thrown across the whole church, the space between it and the back end of the church being the foul-air chamber. This answers the same purpose for the church that the foul-air-gathering ducts (see on Plates IV., V., and VI.) do for the dwelling-house, because all the foul air is here collected together for the shafts (which open in the same chamber) to draw up. This foul-air chamber, as represented in the drawing, is much too large, unless you use furnaces for warming the air; but I was obliged to throw the wall so far forward, in order to *show* the bottom of the foul-air chamber, as well as the bottom of the shafts. If my air-warmers are used, this wall need not be built more than four feet from the back foundations or basement-wall of the church, or from the front of the foul-air shafts. If it be desirable to have the entrance to the basement under the chancel, and from the *back* end of the church, instead of under the vestibule or front end, a passage may be made through the centre of the foul-air chamber, making, of course, good air-tight walls on both sides, and allow not a crevice to be open, to admit any of the external air into the foul-air chamber on either side of the passage. The joists, as you perceive, lie their whole width on the *top* of this wall, so that the air, as indicated by the arrows, has the space of the whole depth of the joists, and the whole width of the building, to flow over the wall into the foul-air chamber. After it gets here, the two apertures, S, into the shafts have a full "swing" at the whole body of foul air.

Should there be no basement under the church, the whole thing is much simplified. The joists may be laid flush with the top of the truss-beams. The floors of the aisles are now laid directly upon them, but the floors of the pews raised four inches above the aisle-floors, the same as in the drawing. The whole space between the bottoms of the joists and the top of the ground, which must not be more than one foot, is now the foul-air chamber, and there is no obstruction to the flow of the air after it gets under the bottom of the blocks of pews, to the foul-air shafts. Double windows are indispensable for all churches, as indeed they ought to be for all buildings intended for the habitation of man in our northern latitudes.

If you lay your truss-beams *longitudinally* of the church with a basement, all you have to do is to lay your aisle-floors directly on the joists, and put your four-inch firring across the joists under the blocks of pews. In this case, as in the other, you must mind to lay the ends of the joists *on the top* of the wall, E, leaving the spaces between the joists entirely open, to admit of the free passage of the foul air into the foul-air chamber.

If, for any reason, it be preferred to have the air-warming machine stand in the basement, the operation will be the same. Let the warmed air come through a register placed in the floor directly over it and in front of the chancel; only, for such a building, there must be at least eight feet of air brought in, and flowing at the rate of five feet per second, and your air-warmer or warmers must be powerful enough to bring this quantity up to 140° from a temperature of 10° above zero, on the outside. This quantity, then, will supply with perfectly fresh and sufficiently warmed air a congregation of five hundred persons. Recollect that I am now speaking of a ceiling not more than twenty feet high. If the ceiling be thirty feet high, you will require more than double the quantity of heat. If the effluent or out-going air should be exhausted *above* the floor, and not thus brought *under* the feet of the congregation, no quantity of heat possible communicated to the air proceeding from the air-warmer would make the congregation comfortable. *The cutting off the cold from the feet* is the great secret of church-warming.

In all buildings liable to have crowds of people in them, and especially if artificially lighted, provision for the escape of animal heat must be made in or near the ceiling. This is so easily done, that it needs no remarks from me.

It is in the erection of our churches, more than any other buildings, that we experience the folly of following on the old jog-trot of prejudice which has descended to us from our progenitors. Brains innumerable, since the civilization of man, have been and are still on the rack to discover the best shape and proportions of buildings for conveying sound. Somehow or other this philosopher's stone has never been discovered. Why? Because the inquirers, being so blinded by prejudice, would never deign to look where *nature* points the way. Confine sound to a tube, and a whisper is heard much farther than the loudest sound which ever proceeded from the lungs of any man in the open air. Sound not only is broken and diffused in a large space, but it naturally rises. The aeronaut will hear the sounds of the multitude below at a much greater distance than the multitude can hear him. Nature's laws must be obeyed; you can not contravene them with impunity. All the science in the world will fail to make sounds as audible in a high apartment as in a low one, no matter what shape you give it. Why not, then, at once abandon the vain efforts, and bend to your destiny, which is so palpably inevitable?

"Oh! it *looks* so horrid—a low church!" How much more horrid would the fur cap and bear-skin or buffalo coat look worn by a man on the plains of Hindostan, or the mere cotton breech-clout in the snows of North-America! Yet the one is just as sensible as the other—a high ceiling in a cold climate and a fur dress under a tropical sun.

Our old architects were sensible men, as the works they have left behind them sufficiently prove. In the erection of their churches, so far from their having any artificial heat to provide for, the very opposite was necessary, and both coolness and ventilation could be provided for by throwing up the ceilings of their churches to a great height. Thus they adapted their buildings to the circumstances around them. Have our architects done so? Let the millions of elderly people who return home, Sabbath after Sabbath, without hearing what was whispered, lisped, or sung, answer. Let the thousands of promising young clergymen who have dropped into an early grave answer.

Architects would be laughed at if they were to differ from "the old masters," and not one, so far as I can learn, has had the nerve to strike out the sensible course. What tends to perpetuate the evil is the vanity of the clergy. "It is so much more respectable to have a large and especially a high church!" So also with the great men of the neighborhood; and the consequences are what I have stated. As to the poor, who are not able to build churches, high or low, they are never thought of. They "have no pews," nor "furs," to sit out an hour or two in a church where the thermometer marks down in the neighborhood of zero. The fact is, the old and the poor, who would derive the most benefit from religious instruction, are excluded from our churches, and little good is done to those who do attend, all from the wretched system of high churches. They can not be warmed, and it is in vain that you try to warm the *heart* of a man who has cold feet.

A church-ceiling should never be more than fifteen feet (or twenty at most) high. Such a church can be easily warmed in winter, and easily ventilated. The people can hear the minister at any season, and when the poor are invited, by abolishing the pew system, our congregations will soon be doubled, especially during the cold season, as well as the stipends of our ministers increased.

PLATES XXXI. AND XXXII.

PLENUM FRESH-AIR DUCTS.

A, self-acting valves. B, sides or walls of duct. C, bottom of duct. D, joists of main floor of building, if the duct be made of wood. E, floor of hall upon which the air-warmer is to stand. F, top or covering of duct. F, G, H, north-west and east ends of duct. These ducts, as will be perceived, may be constructed of wood or brick, for it matters little whether they are hung to the joists or laid under or upon the ground, and the diagrams or plates are so plain as to need little further remark.

The bringing of the fresh air for the ventilation and warming of a building properly under the machine which warms it, constitutes one of the most important points connected with the whole operation.

Such is the extreme mobility of air, that it is *never* at rest, either outside or inside of a house. There may be a few days or hours in the course of a year in which the movement of the external or outside air can not be *sensibly* felt, yet it does move, and it is important that we take advantage of this motion in the ventilation of a building. I must now request you to examine Plates XXXI. and XXXII., which I call the "*plenum fresh-air duct*." They are both the same, only one is made of brick and the other of wood. From a misapprehension of my draughtsman, the plenum fresh-air ducts, to which I have drawn your attention, are reversed, and in order to get a clear understanding of them at once, should be viewed from the north. The references will fully explain how they are to be constructed, but my object just now is to show the utility of this duct in taking advantage of the plenum or full movement of the external air, so that whatever quarter the body of air may come from, the pressure will *always* be *upon* the bottom of the air-warming machine, and consequently through its apertures into the building.

We have so few south winds in this latitude, and especially in winter, that I have omitted making direct provision in my duct for air from this quarter, as the three ends E, W, and N, meaning east, west, and north, will be quite sufficient. Our prevailing winds here, in cold weather, are northerly and easterly and westerly, so that it is quite sufficient to provide for these three points of the compass.

Now a bare inspection of this part of the lungs of a building, if placed under a building, with the F wing north, the G wing west, and the H wing east — let the pressure of the air be from any of the colder points of the compass, it will receive that pressure, and this is all that we require. You perceive that I place valves A at each opening or mouth of the several wings, and these valves open *inward*, so that in case of a west wind, this valve will open, the north and east valves will close, and the whole of the pressure will be *up* the rectangular erection, D D, over which the air-warming machine is, of course, to stand. These drawings were made with a view to the use of my air-warmer, C, on Plate X., but it is obvious that the same principle may be applied to the common furnaces now in use or any other apparatus. I have in the plates of this plenum fresh air-duct hung the self-acting valves in three different ways, either of which modes will probably be equally efficient. I have, however, most frequently followed the hanging of them on a central post, as at H, and which is more clearly exemplified in my car-receiving cap, Plate XLIV. These valves must be made of the very lightest material, (I use tin,) hung with easy-going hinges, and, of course, perfectly plumb, so that a man's breath will move them. In general it may not be necessary to make all three of these wings. If any two will take in the air from the quarter from which the prevailing winds blow in winter, the third may be omitted; indeed, I have, in Canada, made a single one to the north answer very well, but I would never recommend you to risk less than two of them to be used. In some cases a single one run quite across the building with valves at both ends will do. If your building be so situated that you have but one quarter from which you can bring the air, then bring it from that from which the prevailing winds blow *in winter*, and add as many feet to the top of your foul-air shafts or chimneys as possible—to make up by exhausting power what you may fail in propelling. It may not be more than one day in a hundred that all these precautions will be necessary, but it is best to neglect none.

PLATE XXXIII.

OPEN CORNICE AND SMOKE-PIPE REGISTER.

This plan of providing for the warmed air to flow into the different apartments from the hall, may be adopted instead of by the fan-register, F, Fig. 3, Plate III., only it must, of course, be provided for when the house is in the process of building. An aperture must be left all along that side of the room next to the hall, of sufficient size to admit the necessary quantity. For an ordinary-sized room a two-inch opening will be wide enough. This aperture can easily be covered from sight by the stucco-work. It is a very neat way of admitting the air, as a stranger would be at a loss to tell where the warmth came from, and the closing of the draft up the chimney or foul-air shaft is just as effectual in preventing the air coming into the room when it is not needed, as if the cornice-opening itself were closed.

Fig. 2 is intended to admit the warmed air into an upper story or into any apartment into which a stove-pipe is intended to go.

PLATE XXXIV.

This plate was supplied, at my request, by a friend, in order that any person who might want a genteel, and yet a cheap cottage, might be enabled to imbibe the idea from these elevations and plan. It would also enable me to show how the whole building could be warmed, and all the cooking and culinary operations carried on by *one fire*. This last part Mr. Tomkins omitted, but I will now endeavor, as far as it can be done without diagrams, to supply the deficiency.

Plates LIII and LIV. The No. 3 as well as the No. 5 combined are both *cooking*-stoves as well as air-warmers. The No. 3 is plenty large enough for this cottage, and the No. 5, having three boiler-holes, would be sufficiently large for a house three times the size of this one, and a large family. There is no cellar or basement intended to be under the building, but the joists to be a few inches above the ground, to allow of a free flow of air under the whole house. The chimney must be at least thirty feet in height, and have a flue of about two feet, and be opened under the floor. Then put registers in the floor of each principal room required to be warmed, and the exhaustion is provided for. The stove will, of course, stand in the kitchen, and be set up as represented in Plate LIV. A tin tube is then fastened on the end of the cylinder or cylinders which warm the air, and which you see at the front end of the stove, and is carried up over the door which leads into the dining-room in this cheap cottage, but in a large building, with the No. 5, into the hall. Thus a desideratum is supplied, and your cooking and other culinary operations performed, for about two thirds of the year with *one* fire. The oven is made of tin, and is for all purposes of roasting and baking, superior to the common cooking-stove oven; the No. 5 will admit of a much larger one if required.

PLATES LII. AND LIII.

AIR-WARMERS AND AIR-WARMING STOVES.

The distinction between these machines arises from their difference of construction and difference of action. The air-warmers are those whose side plates are *double*, and have little or no direct radiating power, except that which is carried off by the effluent air, which comes from them in three streams, one from the double side plates on each side, and one from the trunk or cylinder. Such are the No. 1 and No. 2, Plate LII., and the car-warmer on Plate LIII.

The air-warming stoves are those which, besides being a common stove to all intents and purposes, warm fresh air by means of a cylinder or cylinders running lengthwise through them. Such are the No. 3, No. 4, and No. 5, on Plate LIII. These by their construction not only do the work of a common stove, but without any additional expense of fuel, will warm distant apartments, and thus ensure a change of air and consequently ventilation, which can not be effected by any amount of mere radiated heat. Whilst the air-warmers warm apartments upon the ventilating principle only, and are, therefore, better adapted for warming distant apartments, the air-warming stoves will also effect this to a less extent, however, whilst they are more efficacious in one large apartment where more active heat is required, such as the hall of a dwelling, or a school-room.

PLATE LIV.

The most troublesome part connected with putting up these air-warmers, and air-warming stoves, I have found to be the getting the fresh air properly to them in old houses. For the purpose of simplifying this as much as possible, I have got this wood-cut made, which embraces my latest improvements.

I have, until within a year or two, contented myself with one wooden duct to come from a northerly direction, as seen on Plate XIII., and also Fig. 2, Plate XXXIV., but my experience has convinced me that the air, to make an efficient job, should come from at least *two* northerly directions. In this plate I take it from the north and the west. In cities and closely-built towns, the winds and movements of the air are so broken up and disturbed, that you may take the air from any quarter with the prospect of almost equal efficiency of action. Indeed, in the cities and towns, the chimneys are generally taller, and therefore have more exhausting power than those in the rural districts and villages, and this circumstance renders the plenum movement of the air less necessary; however, I advise you, wherever it is practicable, to get the air from two directions and use the self-acting valves Y, Plate XLIV.

B, fresh-air ducts. C, brick erection, (or as I have called it, brick penstock,) from cellar-bottom. D, joists. H, stairway in hall. K, top plate of air-warming stove. L, pedestal or hollow back-end of stove.

M, front-end of cylinders. N, to show that the brick erection must of course be hollow from the bottom of the fresh-air ducts upward, at any rate. O, outside wall of house. P P, outside and inside doors of house. R, floor of hall. S, ground or bottom of cellar. T, space between hall-floor and top of air-duct, caused by the joists. E. W. N. and S., the cardinal points of the compass, in order to show you that I have taken the air from those points of compass from which, as I have elsewhere stated, the prevailing winds blow in winter in Canada.

This diagram is made on a scale of half an inch to the foot; the hall is therefore eight feet wide; the stove stands in the centre and between the stairway, H, and the hall-door. For the west fresh-air duct, I have merely perforated the risers of the front steps in order to supply this fresh-air duct which comes through the foundation-wall of a size 10 x 20 inches, which is also the size of the one coming from the North. The artist, I see, has forgotten to put arrows showing the air to go into the perforations of the steps, as he has done in the case of the north duct.

In case you can not get a brick erection, you must make a wooden box, air-tight, the full size required to open the hall-floor, fasten that air-tight to the floor and joists, and put the ends of the fresh-air ducts, B, into *it*.

I have already given the sizes of the apertures to open the floor for the No. 1 and No. 2 air-warmers. Although the Nos. 3 and 4 pedestals will not carry quite 144 inches, nor the No. 5, 200 inches, yet these are the quantities which I generally bring to them, as it is best always to make sure of a *full* supply.

VENTILATION OF RAILWAY CARRIAGES.

The ventilation of these vehicles, although a mere second thought, consequent upon the study of ventilation generally, as well as the agitation of the subject, has engaged my attention, more or less, for several years. I saw from the first, however, that besides the building or adaptation of the carriage *for* or *to* ventilation, in the same way as that of a building, several most important, and some of them additional requisites were necessary. First. A method by which the ventilating air could be purified. Second. A method by which the passengers could be cooled in warm weather. Third. By which they could be warmed in cold weather. And, Fourth. That all the mechanical arrangements should be *fixtures*. No management which falls short of answering these four requisites can become popular or ought to be tolerated.

To such an extent was the fever of excitement increased, and such the competition for the honor (and it may be for the profit) of accomplishing the object, that it became the universal belief that there were many ways in which it could be done. I saw cars rigged up with tin "ventilators," "deflectors," "reflectors," "injectors," "ejectors," "bow-windows," "water-boxes," distributed over the tops of the cars, "pumps, pulleys, and wheels," "bellows-inclosures," "aprons," and other devices following so close upon each other, and all promising to accomplish the "whole thing," and most of them at a trifling expense, that I made up my mind to "bide my time."

The merits of these "inventions" I need not at this time discuss; each has had its day, and the sight of the tops of our cars and the cart-loads of "ventilators" lying about our dépôts bear most melancholy evidence of the ignorance of those whose business it was to judge of their utility and several merits in the first instance, and car-ventilation generally is about where it began.

Even at this period in the history of railways, competition in the passenger traffic has grown quite strong, and in the very nature of things it must increase; so that, even if moved by no higher motive, railway managers will be obliged, if they entertain a proper sense of their interests and the interests of their stockholders, to exert themselves to make their passenger-cars comfortable.

It is quite a mistake in railway managers to suppose that they get *now* all the travellers that they would get under any circumstances.

There are hundreds and thousands who *will not* travel at midsummer or in midwinter, merely because they can not do so comfortably, whereas railway travelling at both those periods *can* be made *more* comfortable than at any other season of the year.

But there are not wanting managers of railways now, who, independently of any pecuniary consideration, have a sincere desire to render their passengers comfortable, and although these gentlemen never, perhaps, hear of the appreciation of their efforts in this direction, I happen to know that their stockholders are deriving much benefit by the unostentatious labors of these managers. It is those only who, like myself, travel a great deal and mix promiscuously with the passengers, and who make it a business to talk of the various improvements going on about them, who are in a position to give an intelligent opinion of public sentiment in this respect. A single word casually dropped by an utter stranger may sometimes have the effect of causing the passengers, or a considerable number of them, to leave the car at the end of their journey with one mind, either of approbation or disapprobation of the management of the road.

I have no pretension to the science of engineering generally; nor yet to the running-gear of a locomotive or railway-carriage in particular; nor do I wish to be understood as presuming to dictate or in any way to interfere in any of those matters pertaining to the running of a train of cars. Although, during a period of about eight or ten years, in which I have been for the most part engaged in experimenting upon the subject of the ventilation and warming of these carriages, I must necessarily have imbibed a general knowledge of the machinery, yet my chief object has been the amelioration of the uncomfortable condition of the passengers. To this it is that I have directed my unremitting attention for several years; and I trust to some effect.

Notwithstanding the numberless schemes and projects and "inventions" which, within the last ten years, have been promulgated for the accomplishment of this object, and notwithstanding that the mania for the exclusion of dust from railway-carriages has drawn forth the talent of this whole continent, it is strange that the simple expedient of making the carriage an entirely close, air-tight one has never been hit upon! This, surely, must exclude dust. The grand error has been the attempting of too much at once. The cooling of passengers is an entirely different thing, and has nothing to do with the first branch — the keeping the dust out. To accomplish this second branch — the cooling of the passengers — requires an altogether different mechanism; as much so as the running and warming of a carriage. It is quite true that in hot weather the passengers must be kept cool by some means; and the only means which we can command consistently, with regard to due economy, is air. "But," says the objector, "wherever air goes dust will go." Admitted, if *all* the body of air surrounding the car in a hot day is the air that is meant.

It is obvious to all that wherever the great body of air surrounding a train goes, there the dust with which it is mixed will go; all the "deflectors," "reflectors," "bow-windows," "ventilators," and "aprons" in the universe will not prevent it. The dust can never be separated from the air which carries it along except by means of water. Now, we do not propose to separate the dust from *all* the air surrounding a train, because we do not require one millionth part of it for the purpose of cooling the passengers. If the air *already in the car* could be set in rapid motion, this, so far as the mere *cooling* of the passengers goes, would answer every purpose; but this, in so contracted a space as the inside of a passenger-car, is impracticable by any natural means. But if it were otherwise, the most important part — the ventilation*— would be left out. A very small quantity, indeed, of air — less than two feet — properly distributed within an air-tight passenger-car, will effect the whole object of keeping a car full of passengers cool and perfectly quiet. Surely we can find means in and about a railway-carriage to purify so small a quantity of air as would flow through a pipe 1 x 2 feet in size. To this the objector may again say, and very philosophically argue: "This is an impossibility, inasmuch as you can not by its own action put a square foot, or any other quantity of air, *into* an air-tight box, unless you take exactly the same quantity and at the same instant out of that box."

When I suggested a "close carriage" to keep out the dust, I meant "close" so far as the passengers are in any way concerned. For, although there is a passage which draws off the air as fast as it comes in, yet the passengers can neither *see* nor *feel* any of its direct effects except as it relieves their respiration, and gives them a cooling and pleasurable sensation.

As the different mechanical fixtures and arrangements (all of which are perfectly simple and easy of construction) are fully explained in the following pages, and as their relation in detail will fully appear when the mechanic ventilates one car, I need not prolong these preliminary observations further than to state, generally, that when the combination of the several parts is completed, the air, including dust, cinders and all other foreign matter, is received at the top of the car, and is propelled by the force exerted by the motion of the carriage through the atmosphere, down flues into a large flat and light wooden pan having a barrel or two of water in it, and lying about an inch deep, and presenting a surface of from one hundred and forty to one hundred and eighty superficial feet; over the *whole* of which surface, by proper arrangement, the air is compelled to roll. This flat pan I call a water-tank. The air then purified is, by the same pressure which, when the car is in motion, is always exerted upon the receiving-cap upon the top of the car, forced up the duct situated exactly in front of the duct which carries it down on both sides of the car, and is from thence distributed, at a velocity of ten feet per second, just over the passengers' heads; setting in motion all the air within the car, but especially that body immediately surrounding the *heads* of the passengers. Thus, the passengers sit and breathe in an atmosphere but little less pure than that in a garden; the car being filled and emptied about every four minutes, and every one *fanned!*

Before I close these preliminary observations upon the rendering of railway-carriages comfortable for passengers, I will take the liberty of making a few suggestions upon improvements which may be made upon our present unventilated cars, and car-*builders* will be good enough, perhaps, to take a note of them.

Every railway traveller, in the winter season, must have experienced, immediately after taking his seat and the car is in motion, an unwonted coldness creeping up from his feet to his knees. This is caused by a current of air sweeping along under the whole row of seats, and which must naturally cause this disagreeable sensation. All that is required to obviate this inconvenience, is to put partitions under a few of the seats, and thus prevent a circulation of air under the whole row. The turning foot-boards at present in use, if fitted up so as to make a whole board of them, may be made to answer the purpose well. Alter their hanging if necessary, so as that when one side is pressed down to the floor, the other side will strike the under side of the seat; thus the foot-board is at once converted into the necessary partition. Another improvement would be to raise that part of the car-floor upon which the seat stands, say four inches. I have had cars constructed in this way, (with, however, another object in view,) and find these raised floors very popular, especially with the lady portion of the passengers, who, when stepping *up*, feel as if they are "stepping out of the dirt." Another improvement: the car-doors should be hung on the *outer* corner of the door-posts, to open *inward*, of course, as they now do; but when closed, the weather-board at the bottom should project half an inch *over* the outer edge of the threshold. This would prevent a large portion of cinders (which lodge in large quantities, especially in wet weather, upon the threshold) from blowing into the car at every opening of the door. I have seen such large quantities of these cinders upon car-floors, where they dry and become pulverized, under the feet of the passengers, and thus become coal-dust, as absolutely to begrim the faces of passengers, to say nothing of the lungs which they affect, and apparel which they injure or destroy. There should also always be an inch clear space left open between the platform and front sill of the car, for the cinders to blow through. Indeed, I know, from actual experience, that greater evils and injury arise from the effects of cinders within a railway-carriage, than from earth-dust, or any other cause. No lungs can withstand the effects of coal-dust with impunity. Here, too, the raised floor shows its superiority over the flush floor; for though cinders may blow under the door into the car, they can not easily rise out of this sunken aisle upon the elevated floor upon which the feet of the passengers rest; whereas, with our common flush floors, the cinders blow over the whole of it, where they are ground into dust by the constant movement and shuffling of a hundred feet.

I had intended here to have introduced some evidence and observations which I have collected upon the all-important subject of the duties and obligations of railway employees toward the public, and especially to their passengers, as well as the duties of passengers toward each other, and also toward the railway company, who at so much expense are exerting themselves to render them comfortable. "At so much expense," I say. I do not mean *pecuniary* alone, but also mental and physical. The *London Review* asserts that "railway managers, engineers, surveyors, officers, and counsel, as a rule, soon lose their health, if not their lives, in consequence of their brains being overworked." As a corroboration of this, you seldom if ever see, in this country, an old man superintendent of a railway. Indeed, as a general rule, almost all employees of railways, especially engine-drivers, are short-lived. Having not, however, quite as much time on hand as I could wish, I must defer this to another occasion. In the mean time I should like to be informed whether, upon the general principle of law that no man has a right to do another an injury, if such injury accrue either by spitting tobacco-juice over a clean carriage-floor, or throwing his dirty feet upon the cushion of a seat or other furniture, ladies' dresses being obliged to be *dragged* over the one, and frequently come in contact with the other, such passenger's ticket is not voidable?

PLATES XXXVII., XXXVIII., XXXIX., XL., XLI., XLII., XLIII., XLIV., XLV.

VENTILATION OF RAILWAY-CARRIAGES.

I had written out a full and minute description, with all the references to each letter, of each diagram, and instructions in detail, how to make almost every piece of their several parts; but having discovered that railway-managers in general *managed* to procure the very best and most intelligent mechanics going, that the diagrams themselves, with *a general view* of the whole process and its operation, would not only enable such mechanics to ventilate their cars, but that it would enable me very much to diminish and simplify my book, that I determined to reject that whole plan, and adopt the latter course.

Where the diagrams or plates fail to explain themselves, the intelligent mechanic will judge of what is to be done by what is wanted, and I will therefore begin by telling him what is required.

I want six hundred inches of air caught on the top of the centre (or near the centre) of the car. This I do by making and setting the receiving-cap, Plate XLV., at that point upon the roof. Of course, when finished, it has another flange, H, upon the other end; and these two flanges (I make them of sheet-iron) are set fore and aft, one toward one end of the car and the other toward the other end. The frame of this cap you will see in Plate XLIV.

I want the whole of this six hundred inches of air to go down the flues, A, Plates XXXVIII., XXXIX., and XLIII., into the water-tank at the centre of each side, where you see the small lines, on Plate XXXVII. Then, after it has gone over the whole surface of the water, (which I shall inform you more particularly about by and by,) I want this air to come up through the lining under the joists, T, Plates XXXIX. and XL.—being a cross and longitudinal section of a car—and flow horizontally through the flue, D, (which is made by the joists between the lining under them and the car-floor over them,) Plates XXXVIII., XXXIX., and XLIII., and flow up flue B; when this flue being closed at the top, it is pressed through the aperture, C, cut out of the front of the flue, B, and into the box, F, Plates XXXIX. and XLIII. From thence it flows through the distributing-pipes or tubes, D, Plate XL., just over and quite close to the heads of the passengers. In winter, when not over two hundred inches of air are needed, it is warmed by the air-warmer, D, Plate XLVII., and is made to flow out of the car under the feet of the passengers, by being forced into one of the end-seats of the car, B, Plate XLI., as seen going into the wire gauze at F; and after going the whole length of the car in this flue, three inches deep and about three feet wide, goes out of the end of a seat at the other end of the car, and up the foul-air flue, R, and so out under the cap, S.

This, in as few words as possible, is what I want the air to do; and I trust, that by revolving the matter in your mind, and after giving you some of the details and demonstrations and reasons which I shall now proceed to do, you will be enabled, especially if you shall have seen a ventilated car — which I recommend you to do before you strike a stroke — you will not find any insuperable difficulty in ventilating a railway-carriage.

As the tank, and its connection with the flues, Plates XXXVIII., XXXIX., and XLIII., are the most difficult part of the operation, I must be somewhat more particular in its description.

The sides and ends I make of two-inch pine plank. It is from about fourteen to eighteen feet long, and nine feet wide — according as the length and width of the car will allow — and ten inches deep. The centre-piece, L, is made of pine scantling, 4 x 6 inches, and the whole length of the tank, and if necessary three inches longer, R, if required to fasten it up by. This piece is worked into a gutter, two inches wide from end to end; two inches deep at the centre, and running out to nothing at each end. (See dotted line, on Fig. 2.) You have, then, two inches of solid wood left on each side of the gutter, to screw your bottom-boards to; which, as you see, are laid crosswise of the tank. K, K, are two-inch boards near ten inches broad, and the length only as far as the heavy lines go, leaving a space between their ends (when set up on their edge, as you see in the next Plate) and the end-plank of the tank of just one quarter of the width of the tank, (you will know the reason of this by and by.) As this tank is to contain from one to two inches depth of water, it is evident that the whole bottom should be divided into compartments, so as to prevent any sudden and undue surging of the water when the car comes suddenly upon an unlevel grade. This inconvenience I obviate in this way: I make eight or ten pieces of inch-boards, N, just of the length of the width of the inside of the tank, and the shape of the lower part of the cross-section, Fig. 3, two inches deep at the centre, and one inch deep at the ends; this is supposed to be the full depth that the water will ever lie over the whole bottom of the tank. After the bottom-boards of the tank are put on, these are laid in on the inside, about a foot or eighteen inches apart, and screwed on from the under-side. Then, in order to make the *other* sides of these water-compartments, you saddle on to them the four longitudinal, M, pieces, leaving a clear space of half an inch under them from end to end for the water and dirt to pass freely backward and forward from the sides of the tank to the gutter, its whole length. Before these longitudinal, M, pieces are laid down, the tank is divided, widthwise, into four equal parts; the two, K, boards (before spoken of) are to be set up edgewise of course, and saddled on the, N, pieces as before described; so as to leave just one of these parts (the one fourth part of the whole width of the inside of the tank) between it and the side-plank of the tank. The upper edges of these, K, longitudinal pieces are now exactly on a level with the top of the tank all around; and you will observe that when the whole tank comes to be drawn up to the level sheathing under the joists, there will be three complete flues; the middle one will contain and carry exactly the same quantity of air that both of the outside ones will. The air comes into the centre of the tank on each side where you see the four small lines, Plate XXXVII., and is propelled round the ends of the K pieces—or as we may now call them, partitions—as shown by the long arrows; and to allow of this is the reason why these partitions were cut off at each end "just one quarter of the width of the tank." As these, K, pieces are standing up as high as the sides and ends of the tank, they need the support of the brace which you see, K, Fig. 3; and their ends, Plate XXXVIII.

Make the bottom of the tank of as thin stuff as you can well match. Not a single nail should be used in putting the tank together, but all *screws*. Indeed, not a nail should be used at all on any part of the ventilation of a car where a screw can be used.

There are so many ways in which this tank may be fastened up to its place, that I forbear to say more than that this must be all done with wood screws, so that it may be taken down easily at any time. I may also say that I have found it about as easy and cheap a way to run up half-inch iron screws, five through each side-plank of the tank, and four (one at each end of the gutter-piece, and two near the middle) up into the sills and the joists.

Mr. Sutherland, at the head of the Michigan Central car establishment, is the best authority to whom I can refer, for not only this but every other part of the work in the ventilation of a rail-car. He is a first-rate workman, has good judgment, and has withal excellent taste, a qualification indispensable in an overseer of a car-establishment. This gentleman fastens up the sides of the tank with four or five hooks on each side, made from about two-inch wide and half-inch thick bar of iron—the ends, about two inches, turned opposite ways — the one he hooks under the side-plank of the tank, and the other he screws up to the sill.

A good deal, however, is to be done before the tank is ready to be put up. A two-inch plug-hole is to be made in the centre of the gutter. The whole of the tank, inside and out, is to have two or three good coats of paint. You then take a quarter of an inch off the top of the two K pieces, and nail on in its place cotton-batting or other soft material, which when the tank comes to be drawn up to the sheathing or lining under the joists, will insure an air-tight joint between it and the sheathing for the whole length, and thus *all* the air is compelled to go round the *ends* of these, K, partitions.

I think now, after again reading my *general observations*, you will understand the reasons for all this particularity in the use and construction of a water-tank. Its use you will see is to purify the air, which can only be done by means of water. It presents the largest possible surface, with the smallest possible weight of water, and the air containing the dust, cinders, and all other foreign matter is compelled to go over the *whole* surface before it can come out of the tank.

The fresh-air chambers on each side of the car I make sixty inches wide and twelve inches deep, which, allowing two inches for the partition and outside board, will give you two flues of five inches deep each in the clear, so that you see each flue will contain 5 x 60 = 300 inches. It is only the *outside* flue, A, Plates XXXVIII., XXXIX., and XLIII., which lets the air *down* into the water-tank; the other flue which stands in front of it merely takes the air *up* after it has gone over the water and is purified. The outside flue, A, only, is cut down *through* the lining under the joists; the flue B only takes it from *above* the lining, and is kept separate from the flue A by the *partition* in the air-chamber, joining the lining, N, between the two arrows, which show the opposite currents of the air, the A flue downward and the B flue upward, see Plates XXXIX. and XLIII. The lining under the joists of the car, you will have seen, forms a complete covering over the whole tank, and unless it was cut somewhere, no air could ever come out of it; but you will see, both in the cross-section and in the longitudinal sections XXXIX. and XL., that the top-covering of the tank is cut open, at the centre of the tank, T, 20 x 30, letting the whole 600 inches of air up, and it flows 300 inches each way between the joists, toward and up the two flues, B, on each side of the car.

I think now that I have said enough in explanation of the plates XXXVII., XXXVIII., XXXIX., XL., XLIII. We will now turn to the "receiving-cap," Plates XLIV and XLV. I see the artist has made the frame 56 inches long—it should be 60 inches long, so as to make it tally better in its joining with the fresh-air chamber, which is 60—clean down into the water-tank. It should be made broad enough to allow of two tin doors, 20 inches long each, and high enough to allow of the doors to be fifteen inches high, clear of the frame, except about half an inch, which the doors ought to lap on all around. B is a partition through the centre of the receiving-cap. All the C pieces I got out of hard wood 4 x 2 inches, the D pieces 4 x 1. The frame is so made that the bottom side-pieces, D, should stand close, air-tight, upon the car-deck, but the end bottom-pieces, C, at both ends, should be so framed as to leave one inch clear at the centre of the deck, so as to allow the air which flows close to the deck to blow clean through under both ends, and so off the car, as you see represented by the far end of both plates. H, sheet-iron flanges on both ends of the receiving-cap, to be on the top, and two sides, about two feet long, but the bottom only about six inches, and this piece should leave an inch clear of the deck, so that the flow of air close to the deck shall not be interrupted. O is a stop that will keep the door from going clean back to the partition, as that much, say three or four inches, will be required, for the air to get behind and close it when the car goes the other way. Y, the tin doors above alluded to. These doors should be made to shut flat against the frame, and hung so as to swing with the least breath of air. The two sides of this cap from D to D are, of course, joined to the fresh-air duct or chamber on either side, and the spaces must be covered over with wire-gauze of not less than a quarter-inch mesh. Plate XLV. is the frame XLIV. covered. P is the deck of the car. H, the sheet-iron flange. X, the top-covering. W, the covering of R, to join to flue A, Plate XXXIX. Y, the doors. These doors, I see, are represented as being fifteen inches square; the other, 20 x 15, is the right size.

With respect to the distributing-boxes represented by F on Plate XXXIX., and C on Plate XL., the draughtsman has somewhat confused the dimensions. I recommend the opening in front of the air-chamber A, and into the flue B, Plate XL., to be made not less than twenty inches horizontally, and sixteen inches vertically, (it might allow of an easier flow if made one inch larger each way.) The shape of the box to cover this aperture may be similar to that on Plate XXXIX. On the two cars which I ventilated on the Erie Railway, I made the distributing-pipes or tubes, D, with a joint, but on the Michigan Central I made them in one piece, and this is the best way. The ends, which are to go two and a half inches into the side of the box, F, should be made twelve inches vertically, by ten inches horizontally; about two feet long and ten inches square at the outer end; the apertures to receive the large ends of the pipes in the sides of the box, F, should be made one inch larger vertically, so as to allow room to move the outer end of the tube up and down about three or four inches. A pretty stiff piece of India-rubber nailed on the inside of the box, so as to impinge upon that part of the tube which is inside of the box, will make it sufficiently air-tight. A quarter-inch wire bolt through the centre of the wood of the box, and through the sides of the tube into the other side of the aperture, will make the *joint* complete. As, however, there is nothing to keep the two *sides* of the tube from coming together, and so letting the air out *outside* of the tube, I have made use of a piece of gas-pipe, with a button made of stiff brass soldered to each end, the wire being made just long enough so that the two brass buttons would keep the tube pressed hard on both sides against the wood. A ratchet near the outer end of these tubes will keep them wherever they are put.

WINTER VENTILATION.

We will now turn to Plates XLI. and XLII.

These are the two opposite ends of a carriage. The air going into one end of the end-seat, B, Plate XLI., and out at the opposite end-seat, G, Plate XLII.

I need not, I trust, at this day expatiate upon the necessity of the ventilation of our railway-carriages in winter. I had written a long article showing the concentration of filth in which we live within these vehicles in cold weather, and especially at night; but I think it needless to inflict upon the reader any argument to enforce that which it appears to me is of itself so self-evident, and shall therefore dismiss this part of my subject with a few general observations.

An ordinary railway-carriage contains something under three thousand cubic feet of air. Supposing there to be sixty passengers inside of this vehicle, and that by the lungs and the cutaneous and other transpiration, each one contaminates ten cubic feet of this air every minute, it is evident that in about five minutes the whole body within the car will have been contaminated. We can form some estimate, therefore, of the intensity or concentration of this contamination, when at the end of a winter's night, of say fifteen hours, the whole body has been thus rendered filthy five times every hour — seventy-five times during the night! It is useless to point to the "ventilators" through the roof of the car, for not a particle of air can go out of *them* for the whole fifteen hours, except a little puff perhaps at the opening of a door, for I have already explained and proved that no air can leave an air-tight apartment, unless that same quantity be *let into it;* and we all know, from experience, that none of the windows will be allowed to be opened during a winter's night.

The lungs of every adult person take in a pint of air at every breath, and this about twenty times a minute; so that into this small tight box of a carriage there are poured twelve hundred pints of matter every minute, the reeking contents of the lungs and stomachs of these sixty passengers — some of them consumptives, and many others, mayhap, redolent of brandy and tobacco! It is quite bad enough for a person to take in a second time the effluvia from his own lungs and stomach, but how exceedingly disgusting is the idea of taking in the emanations, not only from the stomach and lungs, but other parts of the body, of so promiscuous a crowd, for the space of fifteen hours — the matter becoming more and more putrid every minute!

Is it not a wonderful provision of nature that life can be sustained under such circumstances; and is it not beyond all comprehension that an intelligent community quietly submits to and tolerates such a state of things?

Now, couple with this the sufferings endured from cold feet, which rest all night upon an ice-cold floor, whilst the head is in a bath of human filth, nearly up to a blood heat! But we must proceed with our ventilation.

When cold weather comes, the distributing-boxes with the tubes are taken down and laid by for the next summer's use, and the apertures covered over. A water-box is then made the whole length of the width of the air-chamber, sixty inches, and as broad as the space which you have between the air-chamber and the aisle or passage of the car. When set to its place, you tap the air-chamber and make the same sized aperture into the top of the water-box, as close to one end as you conveniently can, similar to B, Plate XLVI., and thus conduct about from one hundred to one hundred and forty-four inches from the air-chamber into the box. As close to the opposite end of this box as you can, you cut an aperture similar to R, S, and upon or over this aperture you set the stove or air-warmer which you see on Plate XLVII. The water-box is made of two-inch plank, cover and all, water-tight of course, and if you find it necessary, cover the inside with zinc. You lay down and screw to the inside of the bottom four ribs, a cross-section of which you see on Plate L., full size, one inch thick and two inches high. They are shaped as you see here, the bevelled side toward the stove, so that the air flowing over it toward the air-warmer will be caught by the point, and drawn down over it upon the water, as represented by the arrow, and so turn over the whole body of air. There must always be kept water over the bottom of this box, one or one and a half inches deep; but it must never be allowed to go dry. The pieces, R, which you see put under the box, are for the purpose of allowing the warm air in the carriage to flow under and help to keep the water in it from freezing. If it should freeze, however, the top of the ice will be thawed in a few minutes after the fire is started in the air-warmer, the whole bottom of which is exposed to the ice. A small door, T, must be made at the most convenient place in the cover, for putting in water and for taking out the cinders, large quantities of which will be collected in a short time. The smoke-pipe, E, Plate XLVII., may be carried up directly through the fresh-air duct, as at W, Plate XLV., or it may be carried with an elbow, so as to come through the roof at one side of it. The aperture made in the cover of the water-box for the bottom of the stove to cover, must be made one inch smaller all round, of course, and care must be taken to double tin all around it, and let the tin come two or three inches over on the top and lap under the plank. Only put the two front screws, H, Plate XLVII., in until you have heated the stove several times, and then when you put the two back-end screws in, let it be done when the metal is about half hot, or you will break some part of it. The register, M, on the top of the stove should be *wired open*. You see the tube F turns the stream of air, which would be very disagreeable to passengers, upward.

There being only about one foot of air required for winter, I generally take the flanges, H, Plates XLIV. and XLV., off the receiving-cap every fall, and stop up one side of the receiving-cap, but you must be careful to stop the two doors that are *both on the same side* of the division or partition that divides the receiving-cap into two apartments, or you will ruin the whole winter's operation; you will not be able to warm your car at all.

We will now suppose that you have your receiving-cap and flues of your air-chamber and tank complete, and painted with three coats of paint, especially the inside of the flue, A. Having taken out all the seats, you are to make two three-inch deep flues the whole length of the car, and the width up to the aisle, as seen in Plates XLI. and XLII. At the one end, generally where the saloon is situated, and which I call the *exhausting end*, you make two foul-air ducts, R, each of which will carry out one hundred inches of air. You cut out and open the *insides* of these ducts, to near the height of the under side of the bottom of the seat, G, upon which the cushion lies, as you see in Plate XLII.; and the top of the flue which you have made being opened the full size of the width and length of the chair, (as seen here, as also at G, Plate XLI., where the air is received,) you have plenty of room to join the air to this foul-air duct. In making this flue, you must cut round the fresh-air chamber, of course, notwithstanding it somewhat diminishes the quantity. At the receiving end of the car you see the foul air enters the end of the seat through wire-gauze, F; it does just the same on the opposite side, at G, and flowing along, at about the rate of ten feet a second, goes up and out of the foul-air ducts, R, and so out of the cap, S. The chairs being put back upon the top of this flue, all the passengers' feet are resting upon a warmed-air flue, completely cutting off all cold from their feet.

PLATES XLVI., XLVII., XLVIII., XLIX.

VENTILATION OF CARS FOR WINTER ONLY.

What I have hitherto written upon the ventilation of railway-carriages has had relation altogether to what I call the FULL ventilation, that is, for both winter and summer; but as it may happen that railway companies might want to adopt this without the other, I have got the plates from XLVI. to L., inclusive, lithographed.

This is a much less expensive operation, as you need no double floor or expensive receiving-cap, water-tank, or fresh-air chamber or ducts. This will, for the first cost, be a little dearer than the car heating by a single stove, though nothing like the expense of the two coal-stoves, which I have seen on some roads. The advantages which this mode of warming possesses over stove-heating are these: you have as thorough and speedy a change of air as in my full ventilation, that is, there will be no air in the car six minutes old; you have an equable temperature throughout the whole car, and you do the work with half the expense of fuel.

The first thing is to make the water-box, the plan of which you see on Plate L., and a perspective view, Fig. 2, Plate XLIX. I make it of two-inch plank, water-tight of course. As in this case there are no fresh-air chambers in the way, the box may fill up the whole space between the side of the car and the aisle, say three feet broad, and it may be four feet long. The partition, F, which you see in the perspective view, is put near the centre of the width of the box, in order to compel the air which comes in on the side opposite to where the stove stands to take a longer circuit over the water, as you will see by the make of the cover and representation by the arrows. The air-warmer may be set next the side of the car, and the fresh-air pipe next the aisle, or the reverse way. The fresh-air pipe, A, Plate XLVII, should be ten or twelve inches diameter, and run up through the roof, and the cap, Plate XLVIII., (which I have made on a much larger scale,) put upon the top of it. This cap is made of tin or sheet-iron, and when a ten-inch diameter fresh-air pipe is used, fourteen inches square. There is a top, B B, bottom, D D, and two sides, B and C C. The two ends, which are left open, are, when set on the deck of the car, "fore and aft." The valve, F, Plate L., slides backward and forward freely upon the two wires, K, which are fastened at the ends, as you see at H, Plate XLVIII. The flanges, E, are turned *in*, so as to prevent the valve from blowing *through*. The wires may be near a quarter inch thick. In order to make the sliding-valve, F, slide over the wires easily as possible, you must solder firmly into it, at Q, two thimbles, about two inches in length, one inch on each side of the tin. This, of course, must be of a capacity half as much larger in diameter as the wire, so as to have it work perfectly easy. The bottom of this cap must be well braced, so as always to maintain its *level* position, as you see this is at G. The loop, I, must be put on both ends on the top, to wire the whole cap to the deck, and this must be strongly and substantially done, as sometimes it is liable to great stress. The openings at the lower end of the flanges, P, are left there to prevent the accumulation of snow, and so prevent the valve, F, from setting *close* to the flange.

When this cap is properly joined to the pipe, A, Plate XLVII., it is easy to perceive that, let the car go which way it will, a body of air equal to the full capacity of the pipe will be driven down into the water-box.

The "ribs," A, B, C, are made the full size of the cross-section on Plate L., and are nailed to the bottom of the box disposed of as seen on Fig. 2, Plate XLIX. Their business is to cause the air on its way to the stove or air-warmer, to turn over. The top of the box must, of course, be supplied with the door T as on Plates XLVII. and XLVI., and the R pieces also under it. From one to two inches depth of water must always be kept in this box for the reasons given in regard to the same box in the instructions on the full ventilation of cars, and to which I refer you for further directions. The whole thing should not, besides the stove, cost over twenty or thirty dollars.

A pretty good *summer* operation may be carried on by making and setting one of these boxes on each side of the car, and setting a hollow tube upon each, five or six feet high, and distributing the air something similar to that which I have described in the full ventilation. But after all, the addition of four or five per cent upon the cost of a car, for its *full* ventilation, is a mere bagatelle when compared with the comfort which your passengers will enjoy beyond that in the unventilated cars for either summer or winter.

With a full ventilation in summer, the passengers sit free from any dust except that which is incident to any apartment, and in fact more free from dust than in our ordinary rooms covered with a woolen carpet and occupied by the same number of people, and every one may be fanned with as pleasant a breeze of pure air as upon any of our lake or river steamers. All the air within the car being changed every four minutes, you are entirely exempt from the smell of the bilge-water, grease, and heat of the engine and kitchen on board our lake and river steamers, as well as the qualmishness always more or less induced by the motion on the lake.

In winter all the air is changed every six minutes; the car is, with the exception of that part in the *immediate* vicinity of the air-warmer, as warm at the ends as in the centre, and every passenger's feet are kept perfectly warm by the warmed air, running at the rate of ten feet a second, under them from one end of the car to the other.

PLATE LI.

This plate is given merely for the purpose of showing, that when you build a chimney with two or more flues in it, as seen at E and F, and on Plates II. and IV., the flues should be run *together* before they get out of the last story; as you see here, the flues C and B join at D. The smoke-pipe, if you have one, should be put in here where you see the dotted circle. From this point upward, although the flues each contain 144 inches up to where they join, the whole flue, A, upward from this need not contain more than one and a half feet or say 216 inches, and this should again be diminished at the very top about one eighth, as you see.

When you meet with an old two or three-story house, having two or more flues both running clear out to the top, put your smoke-pipe in that one which is connected with the largest room, or that which is most used, below, because its exhausting power will be much more than the other.

PLATE I.

FIG. 1.

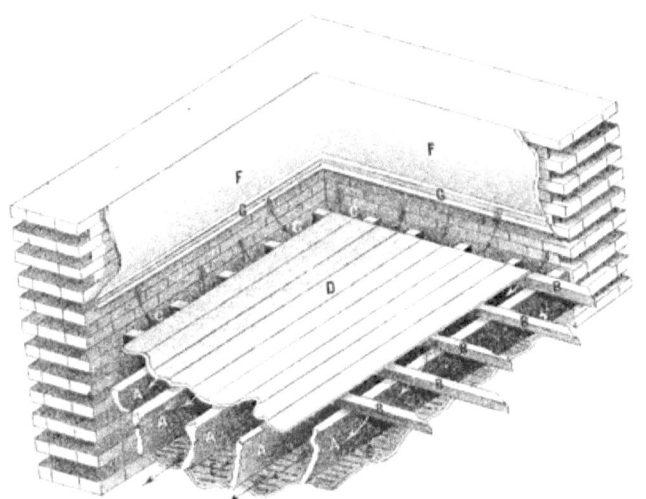

FIG. 2.

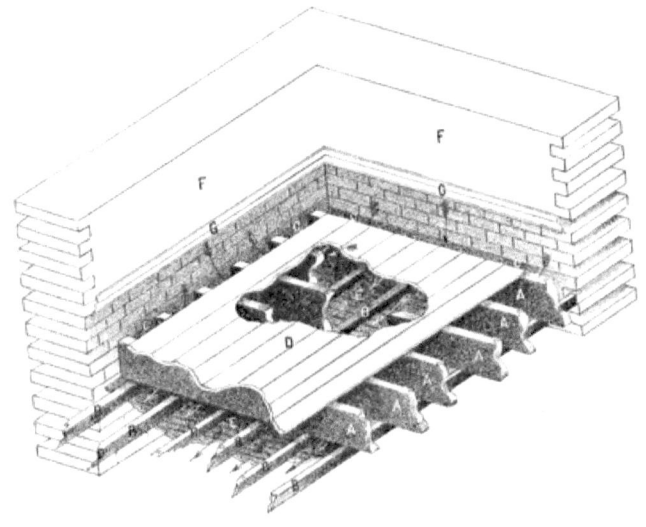

PLATE II.

PLATE III.

FIG. 1.

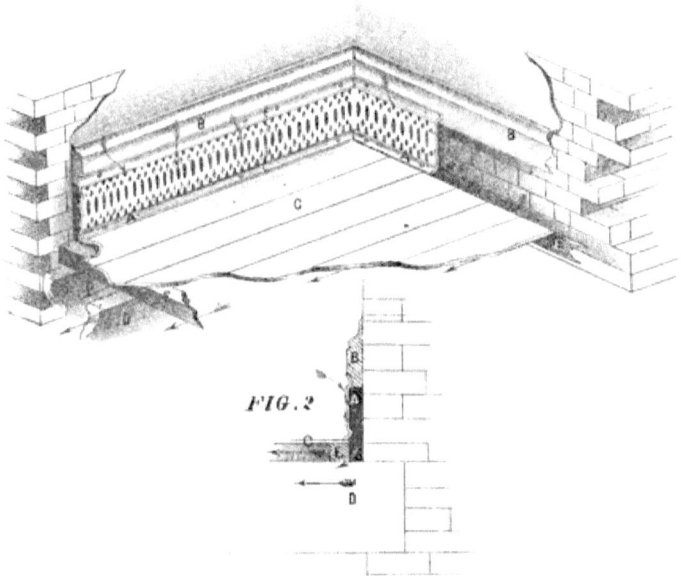

FIG. 2.

FIG. 3.

PLATE IV.

PLATE V.

PLATE VI.

PLATE VII.

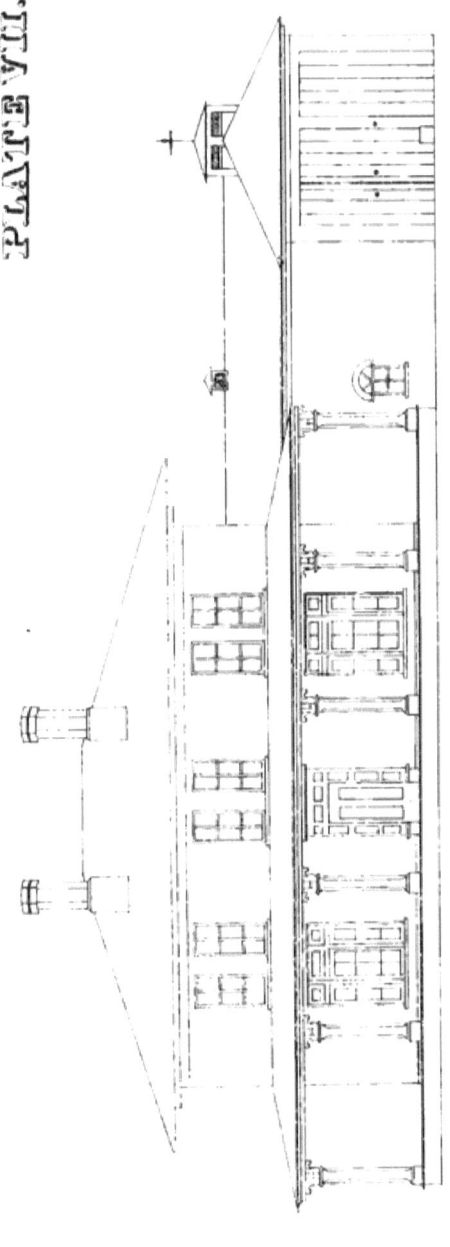

FRONT ELEVATION.

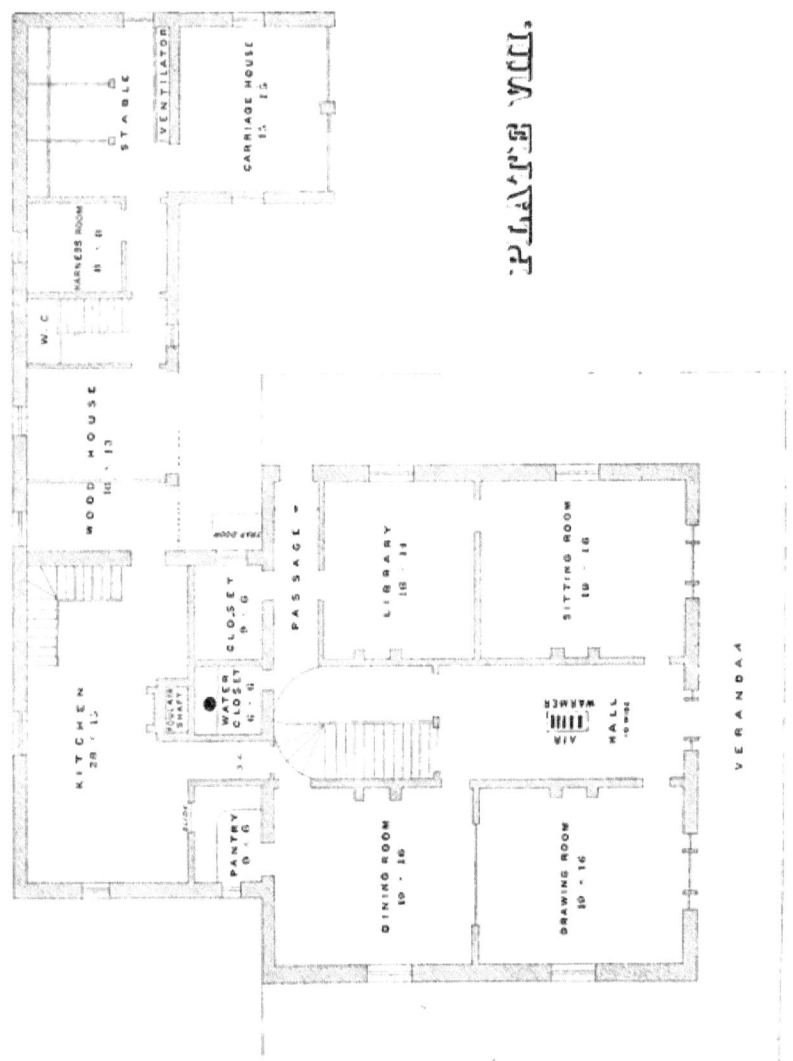

PLATE IX.

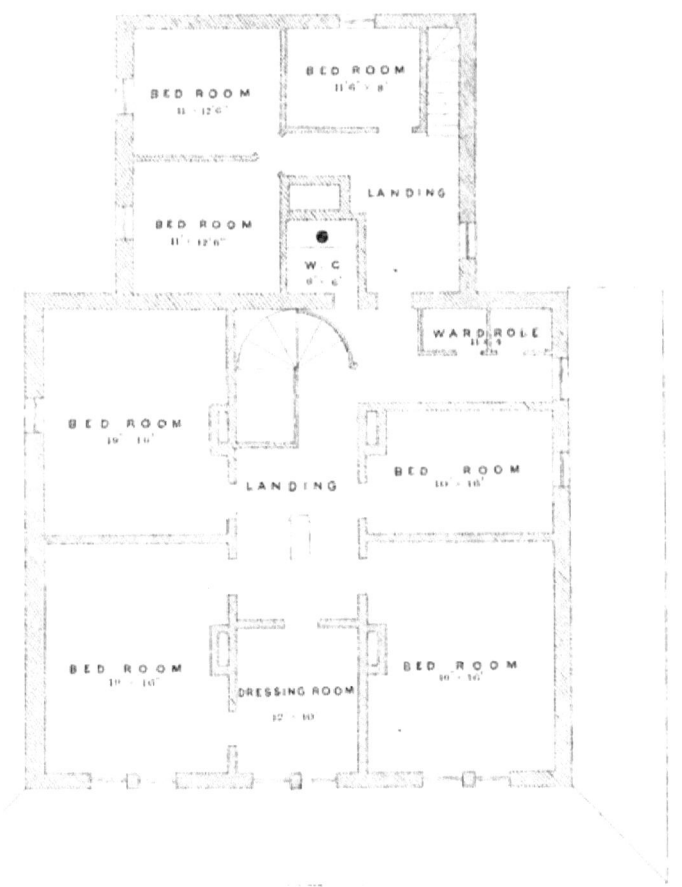

SECOND FLOOR.

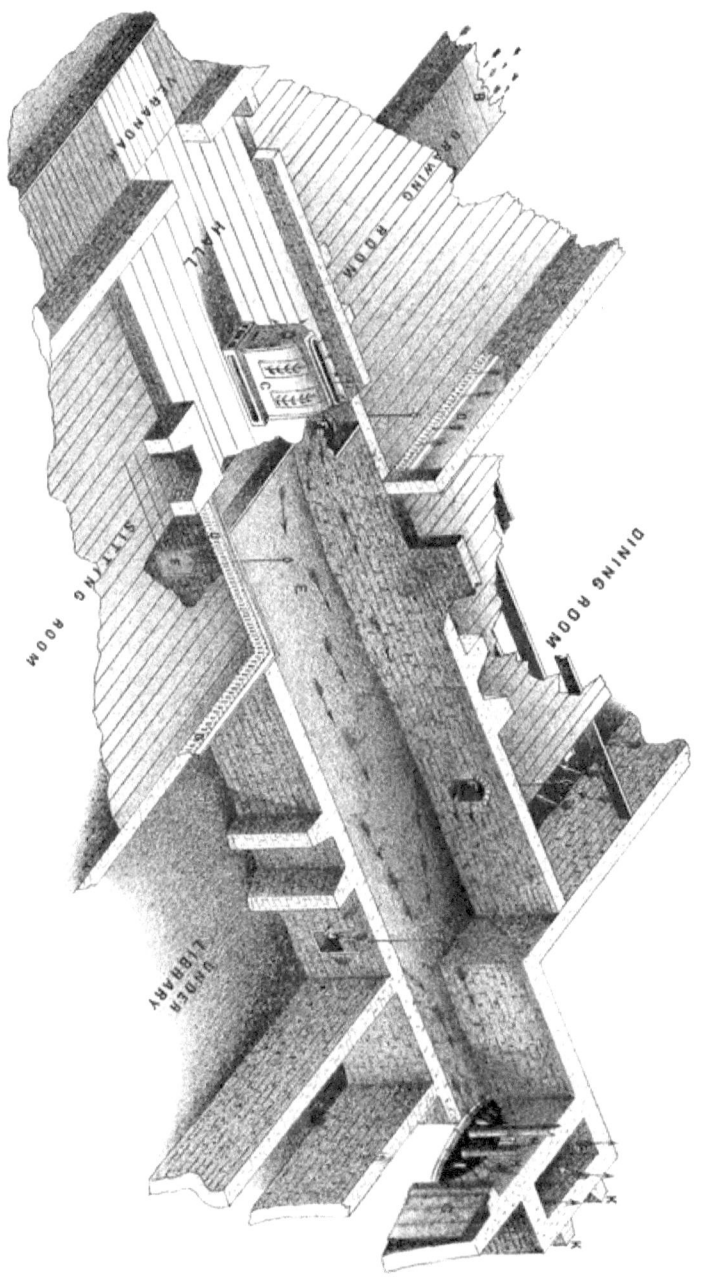

PLATE X.

PLATE XI.

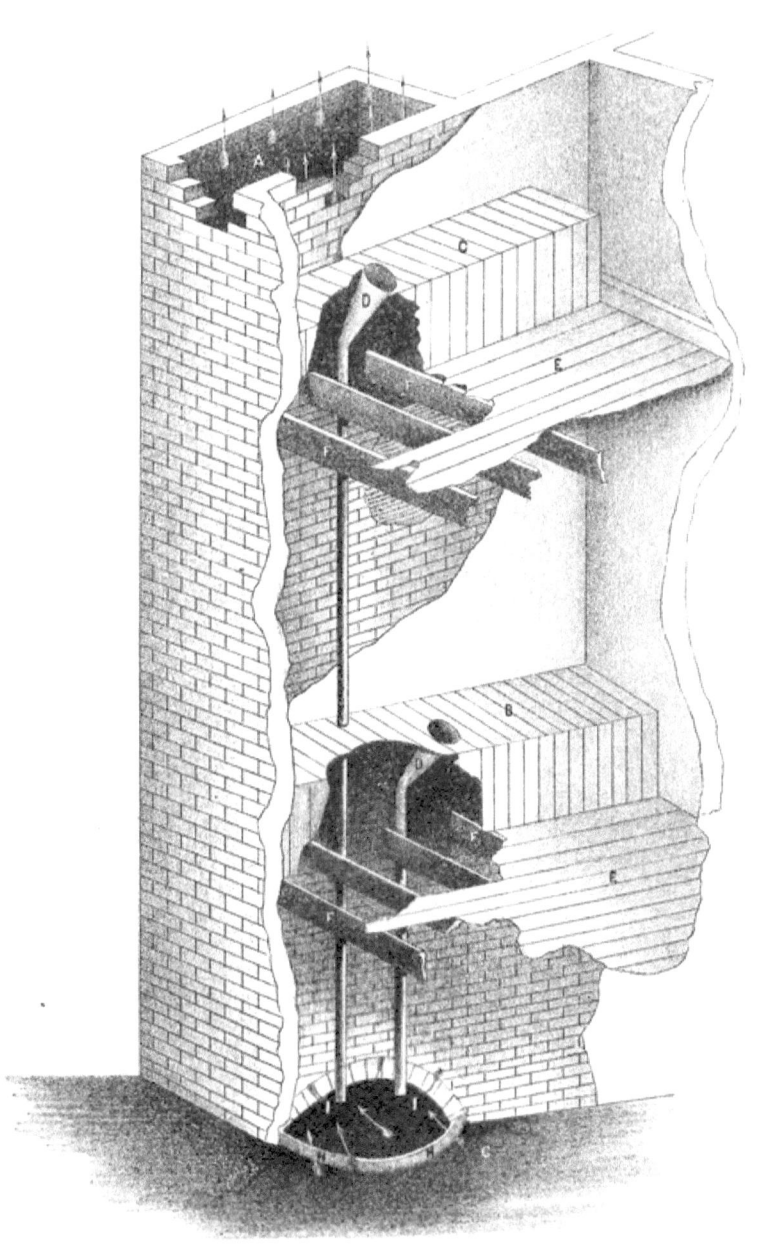

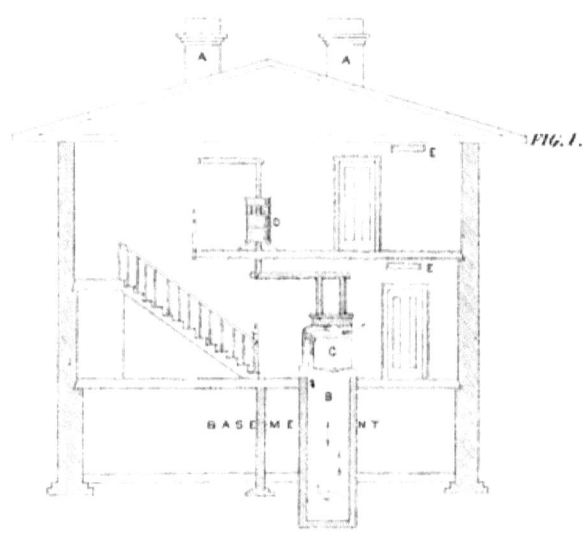

PLATE XII.

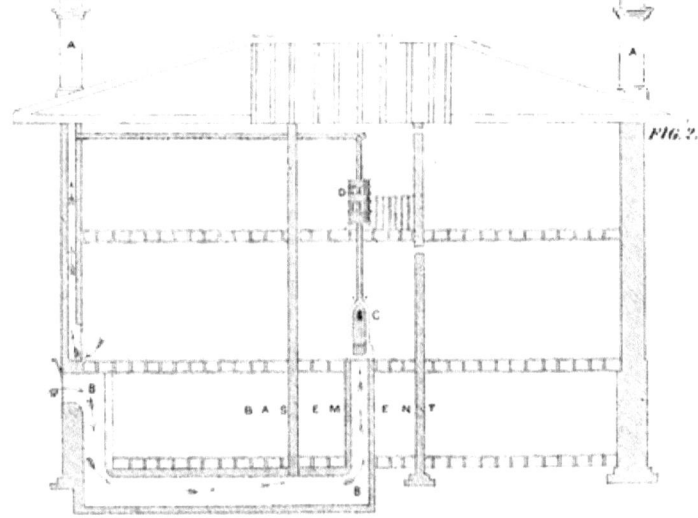

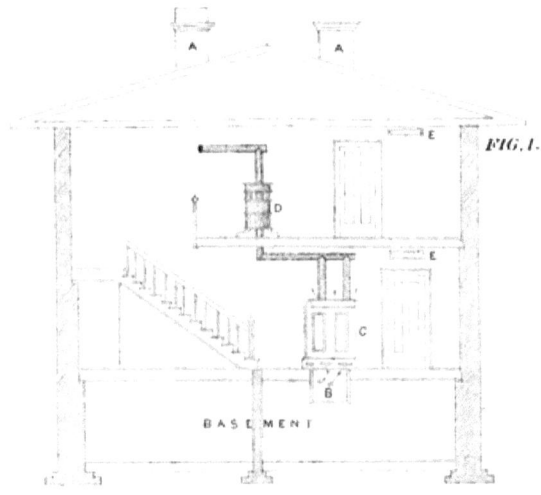

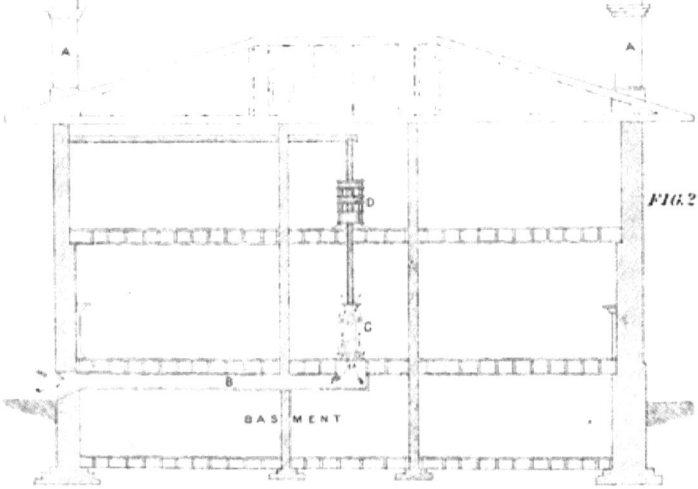

PLATE XIII.

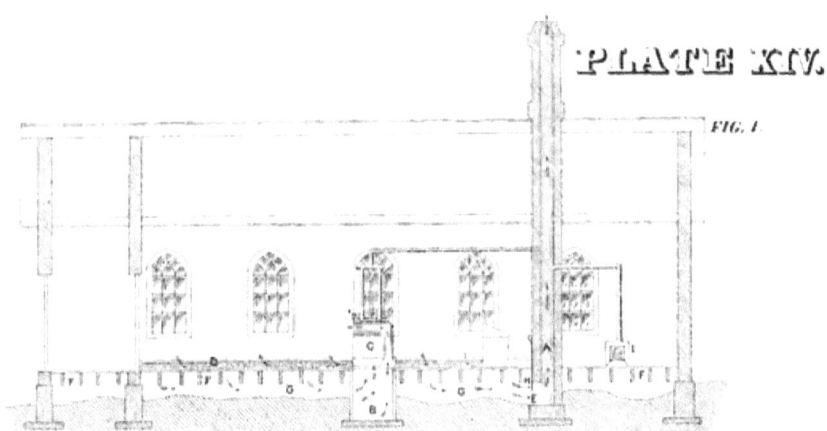

LONGITUDINAL SECTION ON LINE A.B.

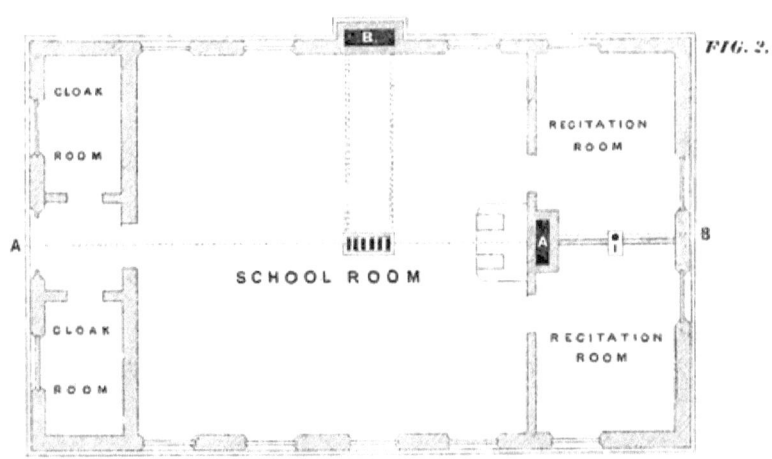

GROUND PLAN

PLATE XV.

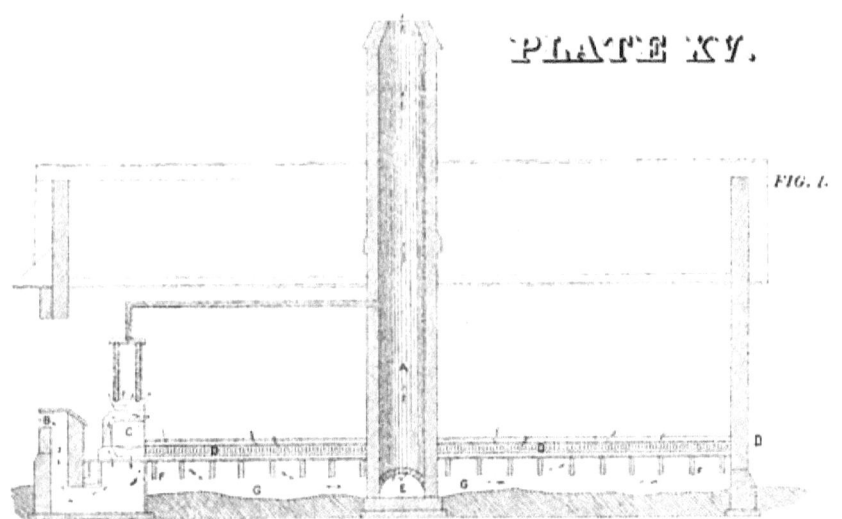

SECTION ON LINE A B

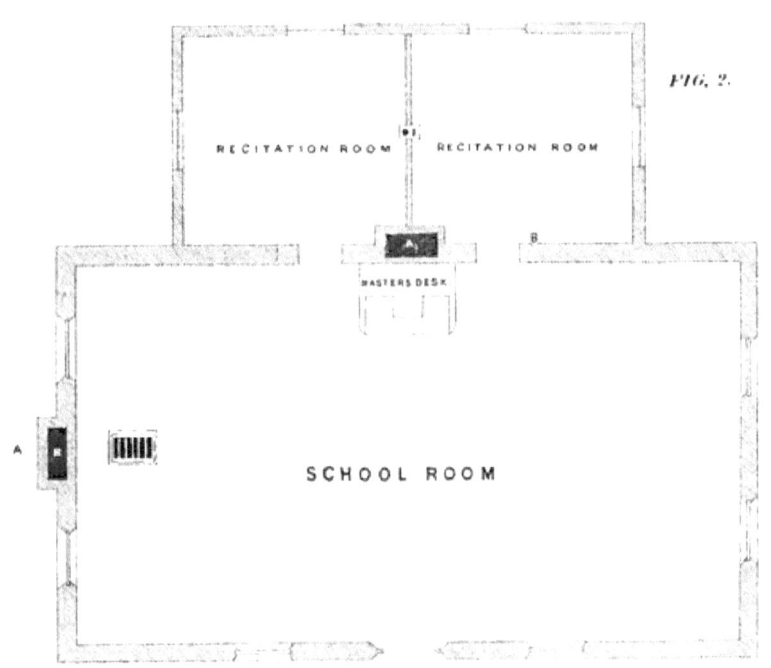

GROUND PLAN.

PLATE XVI.

FIG. 1.

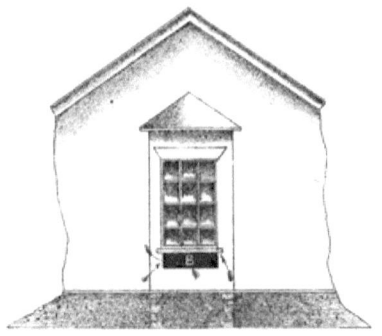

ELEVATION SHOWING OPENING
FOR FRESH AIR UNDER WINDOW.

FIG. 2.

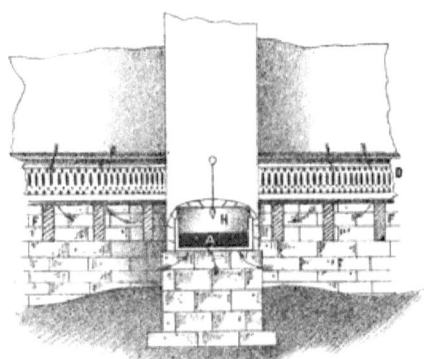

ELEVATION OF FOUL AIR SHAFT
SHOWING OPENING FOR ESCAPE OF
FOUL AIR AND REGULATING VALVE.

PLATE XVII.

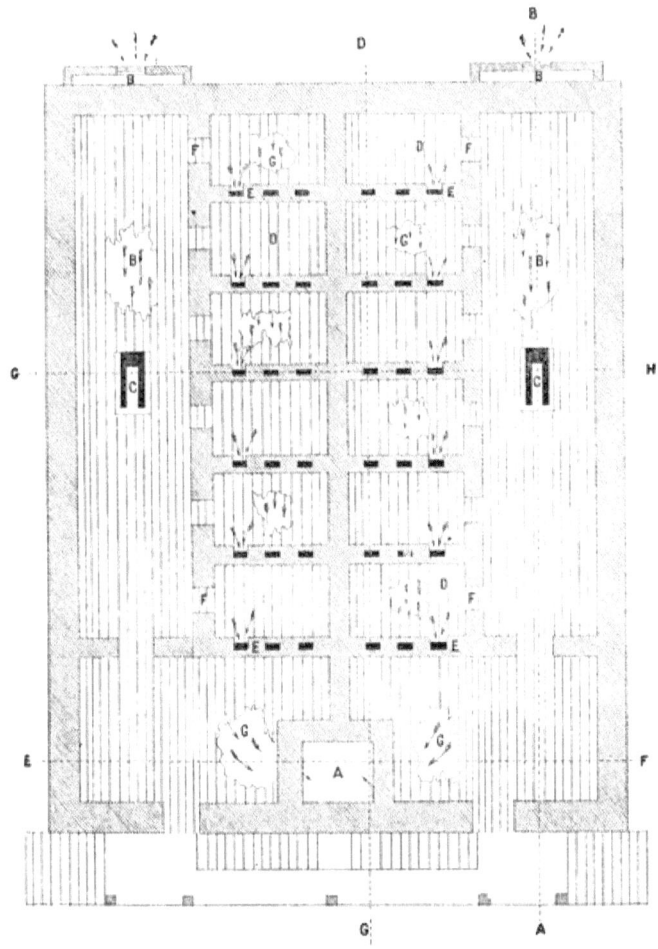

PRINCIPAL FLOOR

PLATE XVIII.

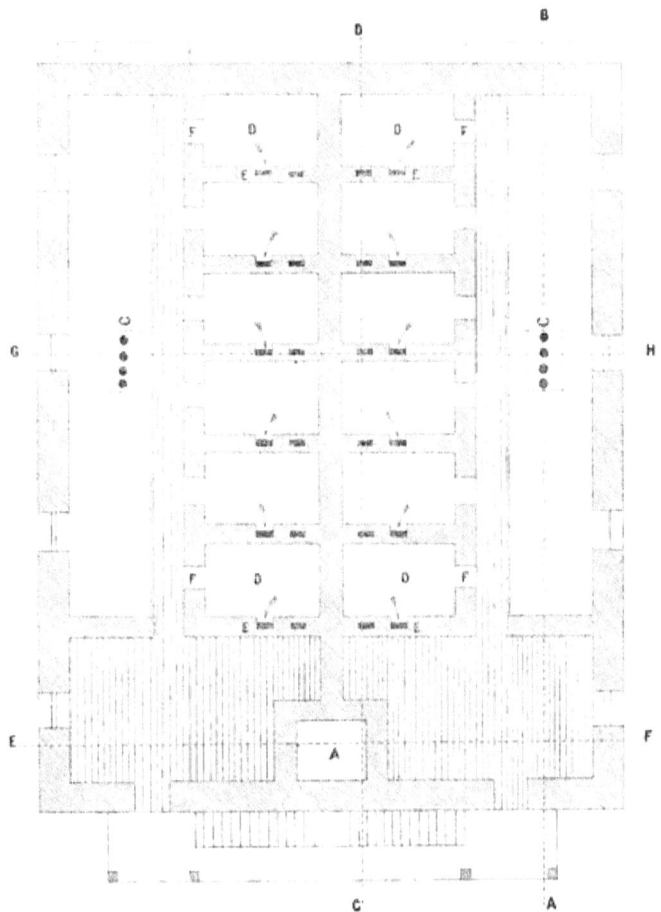

SECOND FLOOR

PLATE XIX.

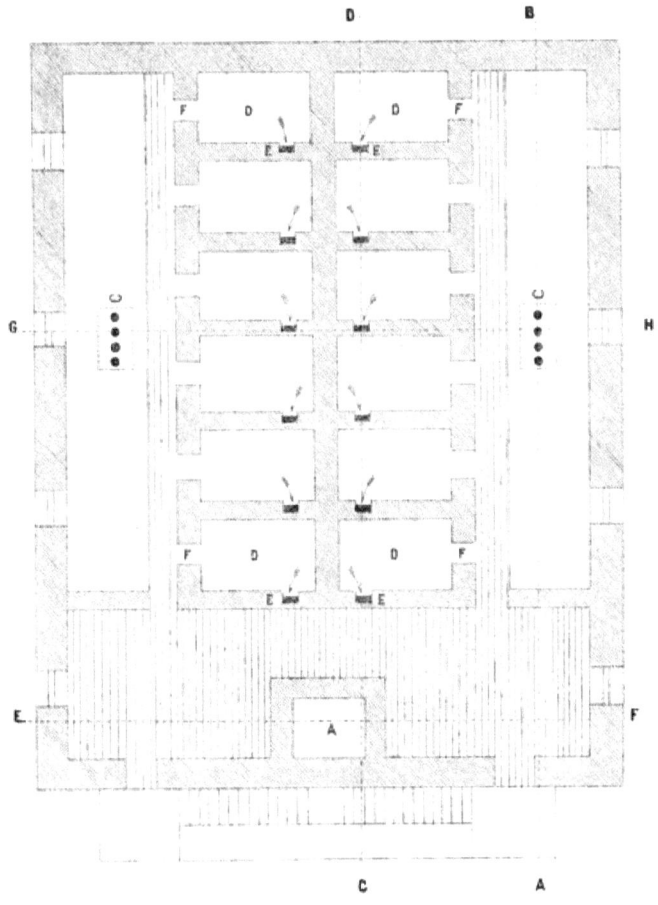

THIRD FLOOR

PLATE XX.

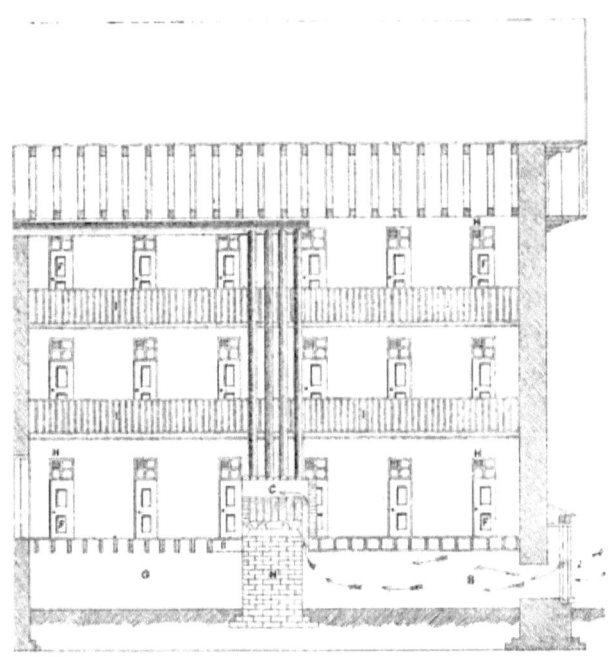

SECTION ON LINE A.B.

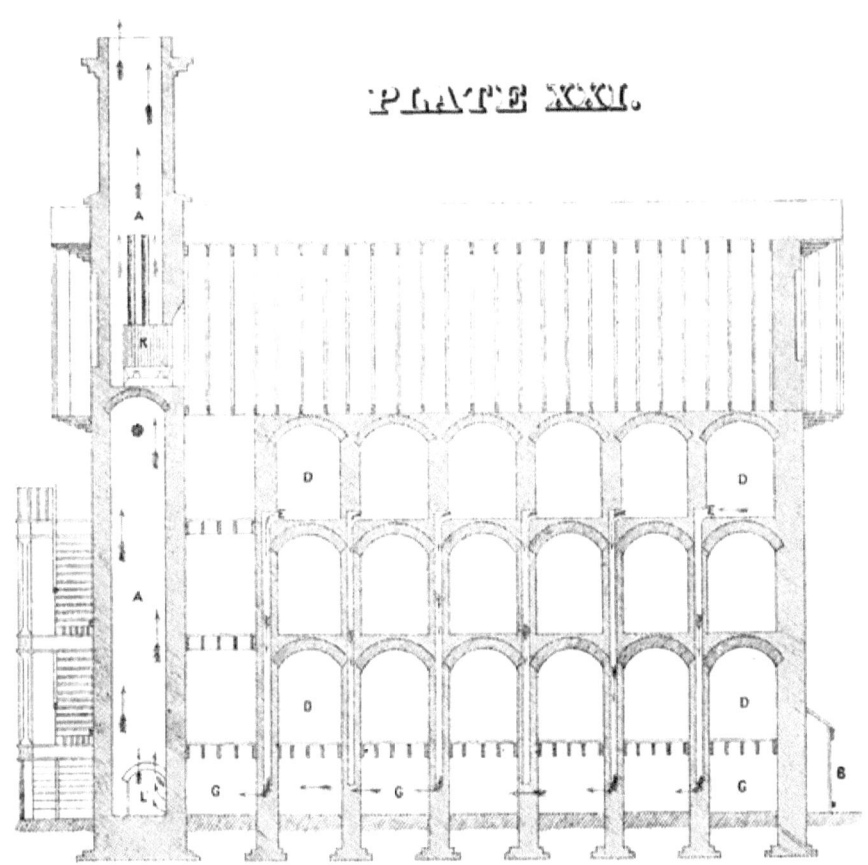

PLATE XXI.

SECTION ON LINE C.D

Scale 12 Ft to 1 inch.

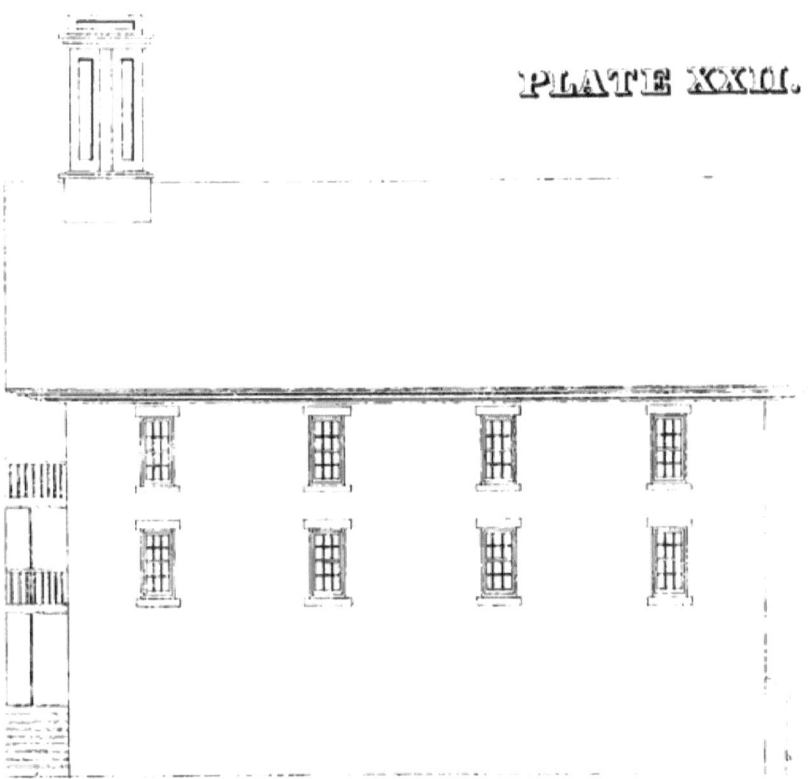

PLATE XXII.

SIDE ELEVATION.

PLATE XXIII.

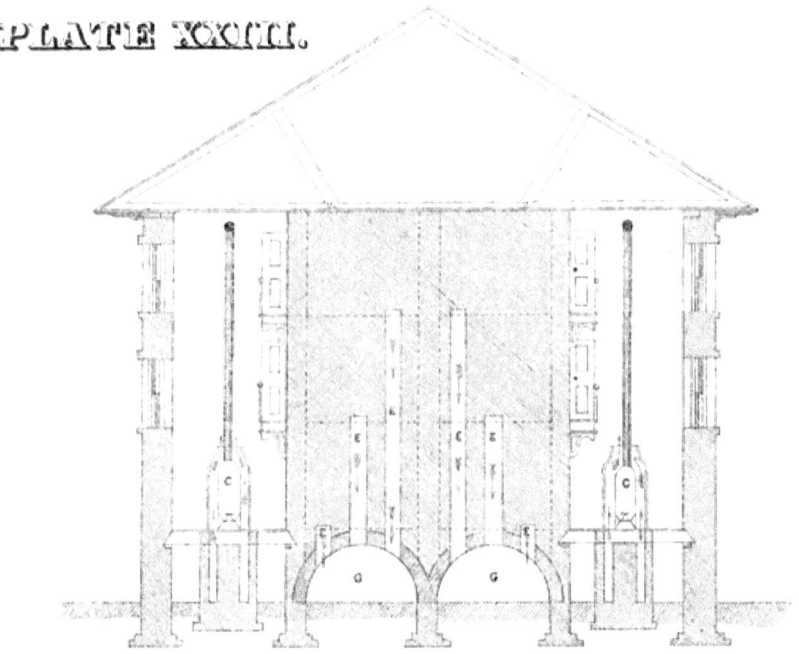

SECTION ON LINE G.H.

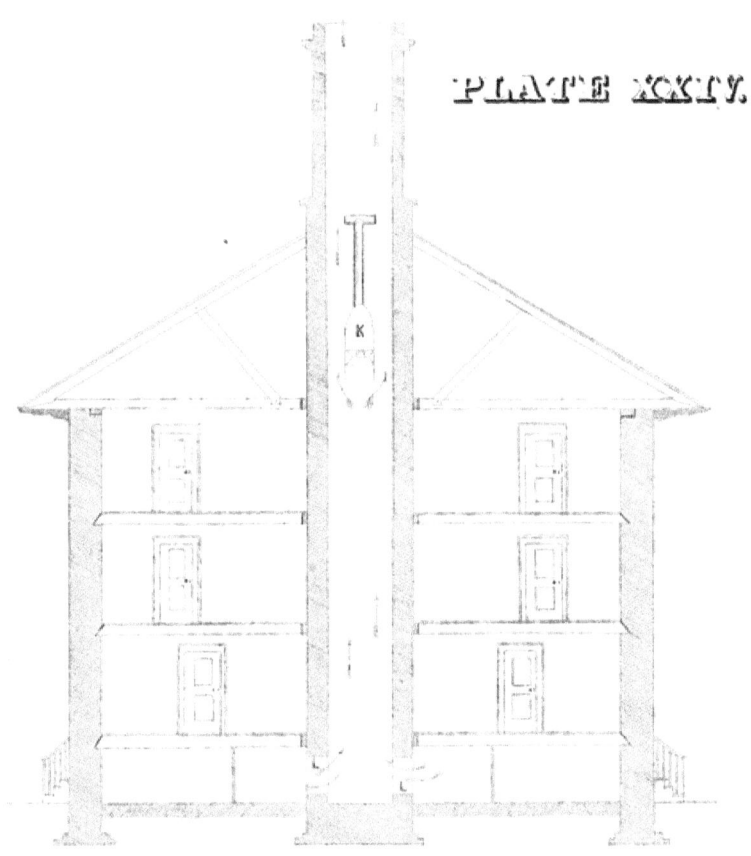

SECTION ON LINE E.E.

PLATE XXIV.

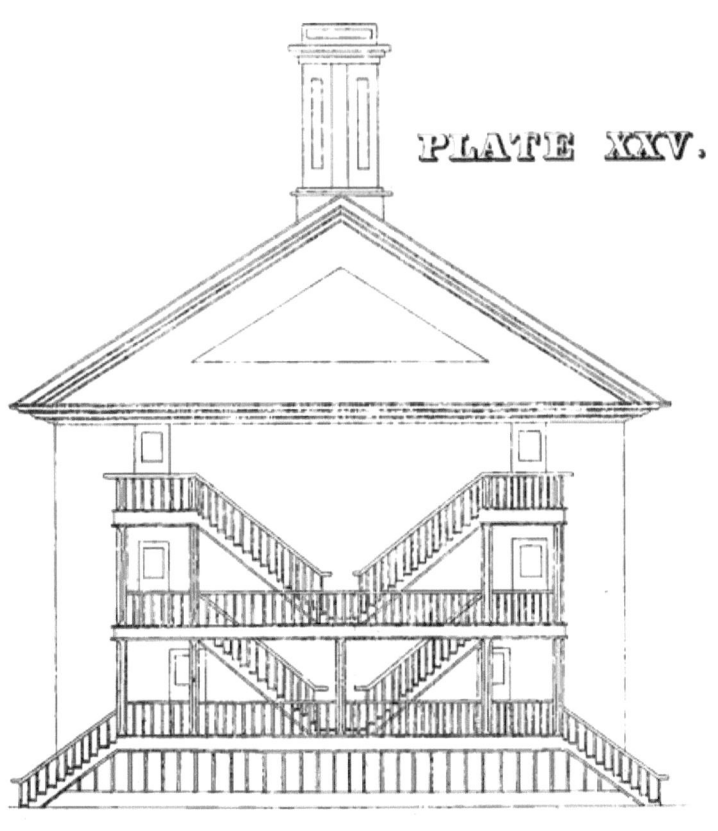

PLATE XXV.

END ELEVATION

PLATE XXVI.

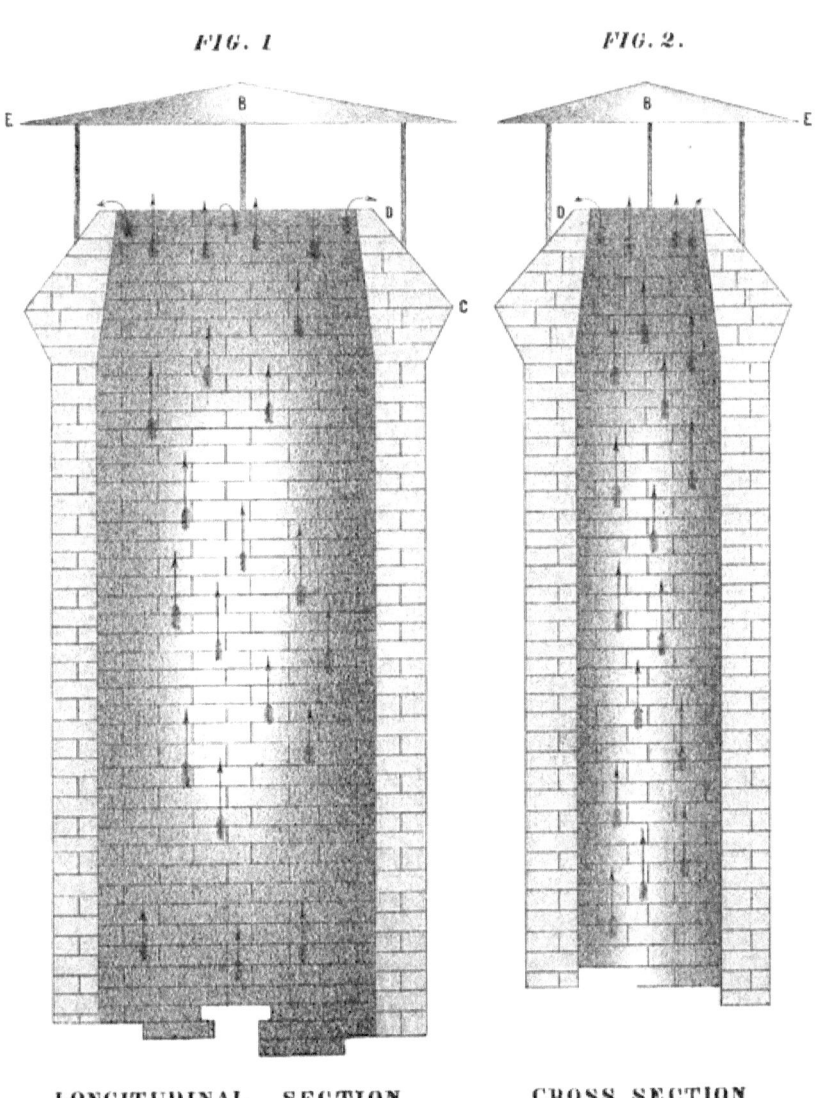

FIG. 1 FIG. 2.

LONGITUDINAL SECTION   CROSS SECTION

Scale 2 feet to 1 inch.

PLATE XXVII.

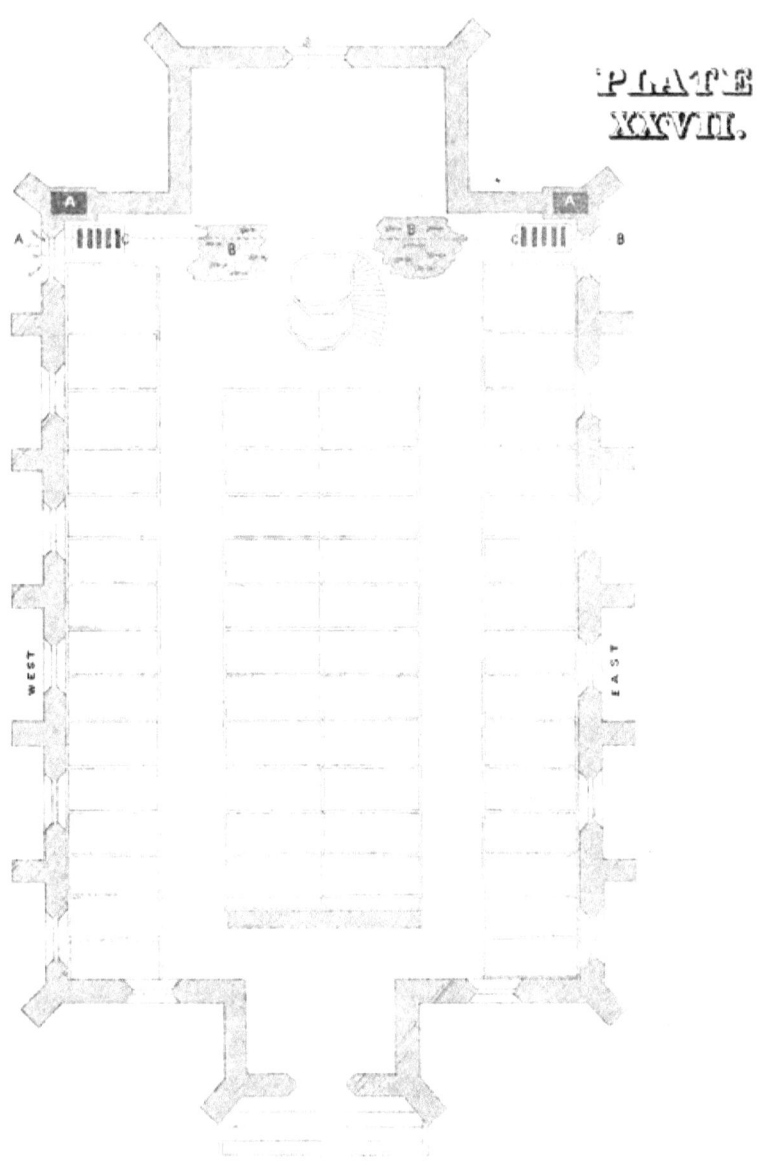

GROUND PLAN
SOUTH

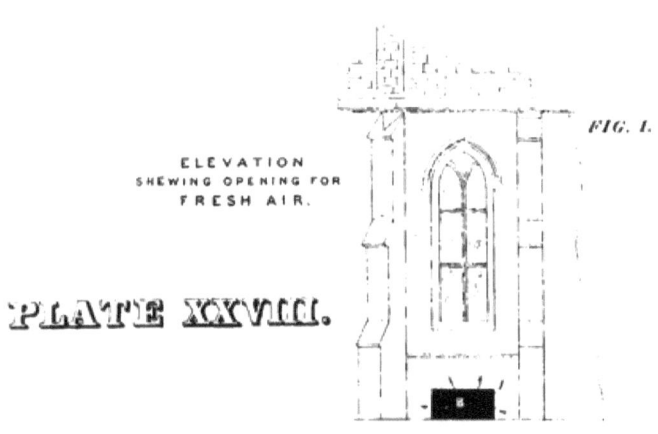

PLATE XXVIII.

ELEVATION
SHEWING OPENING FOR
FRESH AIR.

FIG. 1.

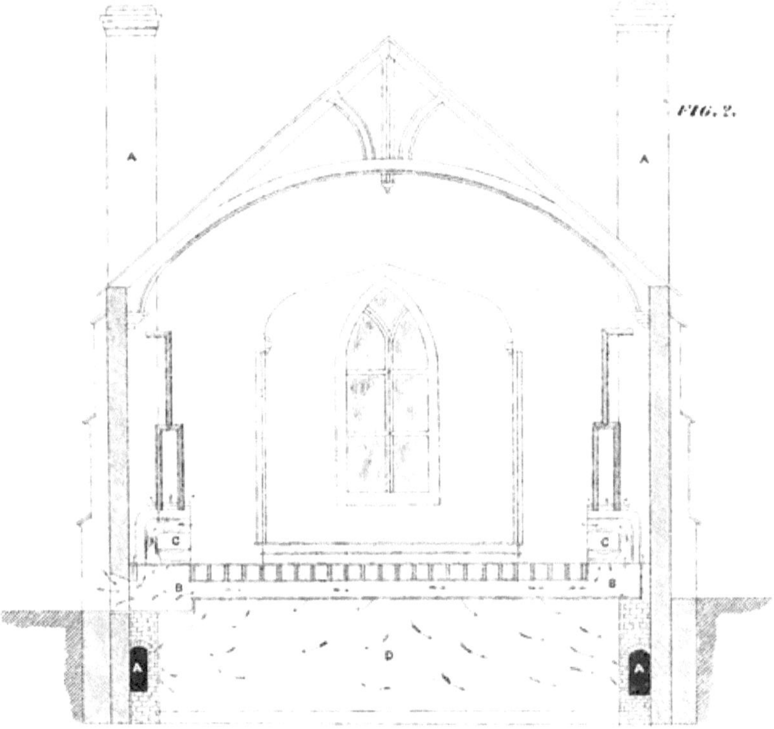
A
A
C
C
B
B
D
A
A

FIG. 2.

SECTION ON LINE AB.

PLATE XXIX.

SOUTH ELEVATION.

CHIMNEY CAP.

FIG. 1.

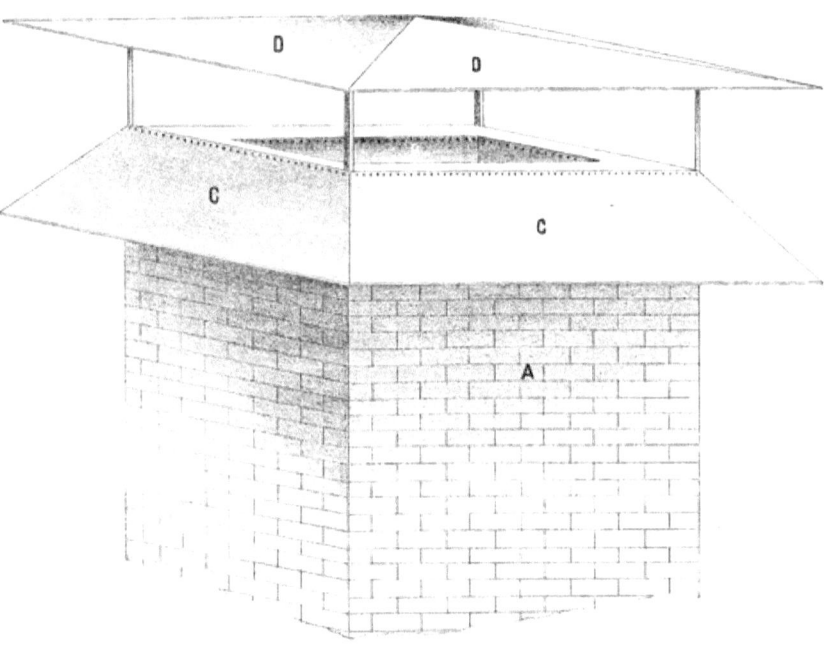

CHIMNEY CAP TURNED UPSIDE DOWN SO AS TO SHOW THAT PART WHICH GOES **INSIDE** OF THE FLUE AS WELL AS OUTSIDE

FIG. 2.

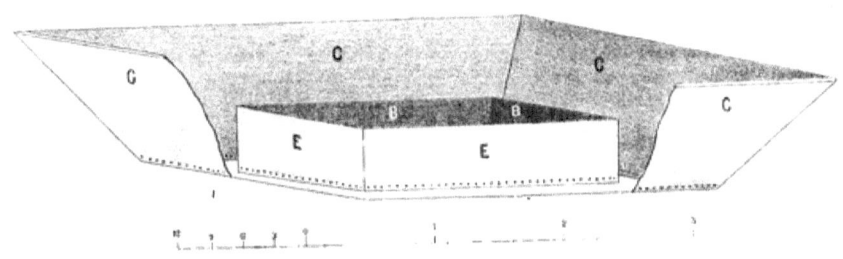

Scale: 2ft. to an inch

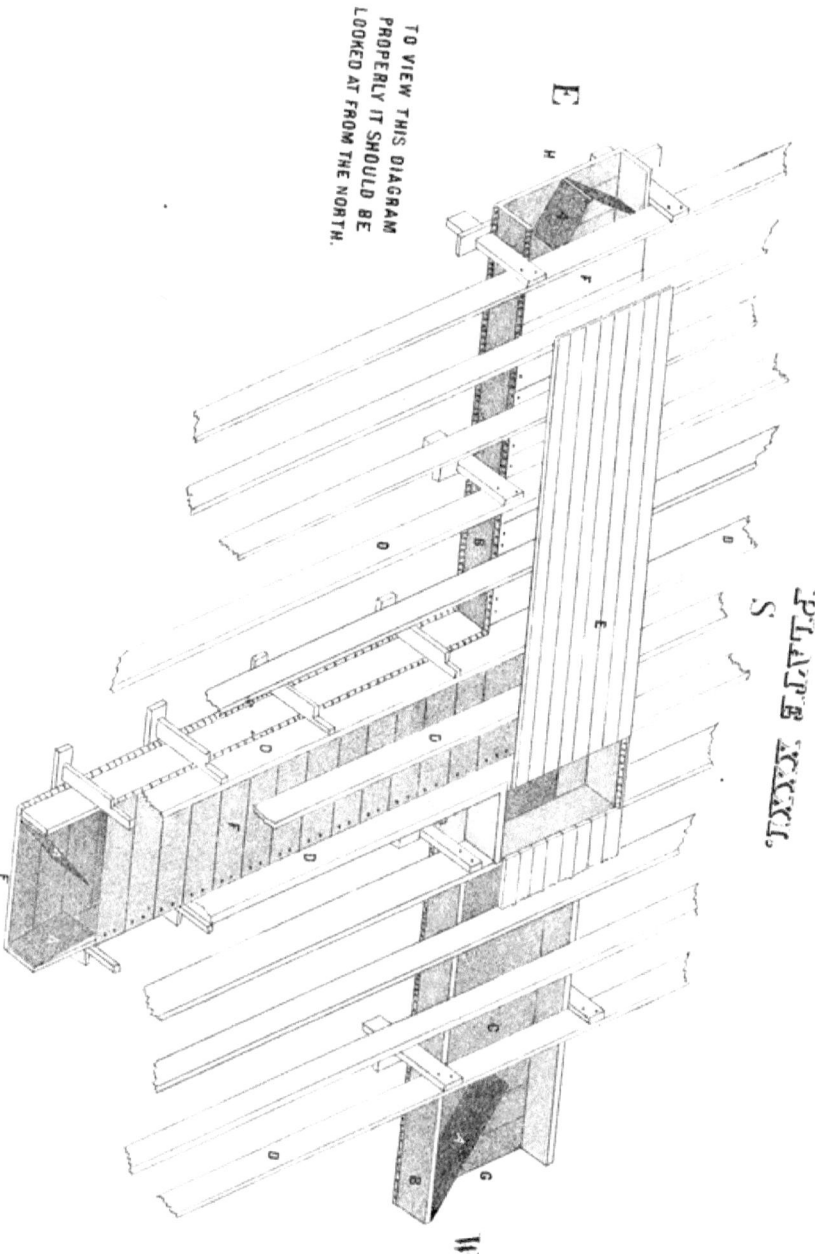

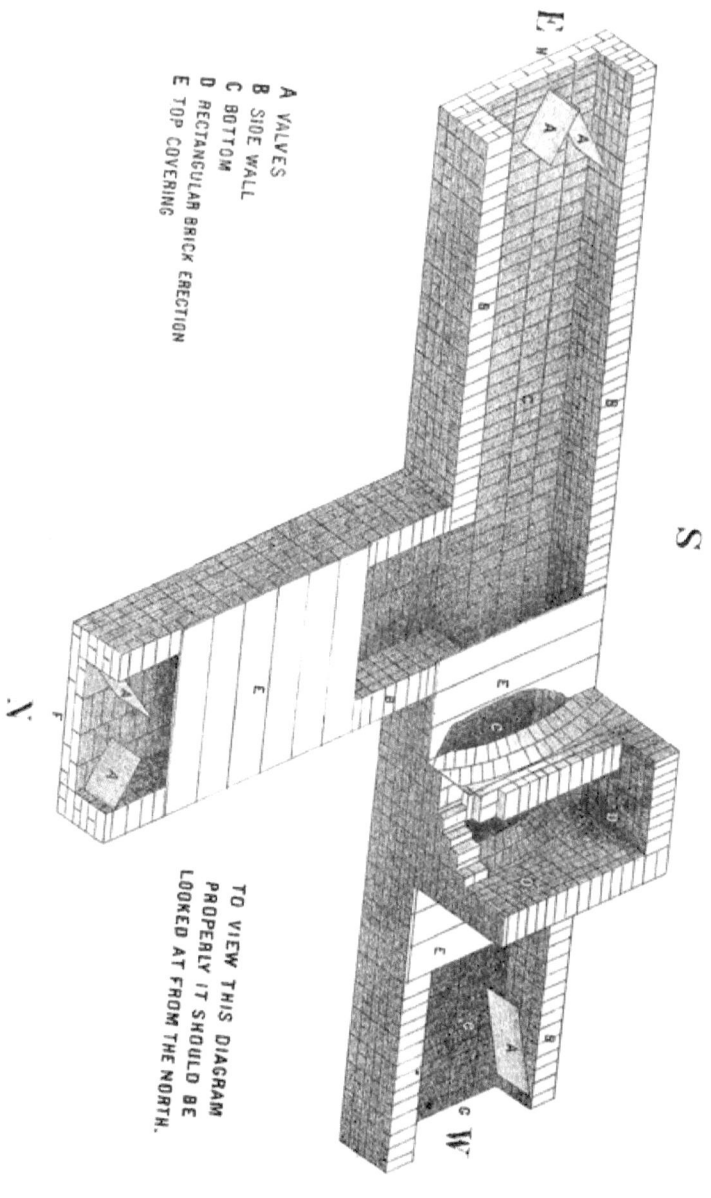

A VALVES
B SIDE WALL
C BOTTOM
D RECTANGULAR BRICK ERECTION
E TOP COVERING

TO VIEW THIS DIAGRAM PROPERLY IT SHOULD BE LOOKED AT FROM THE NORTH.

Scale 2 Ft. to one inch.

FIG. 1.

CORNICE

PLATE XXXIII.

FIG. 2.

REGISTER

PLATE XXXIV.

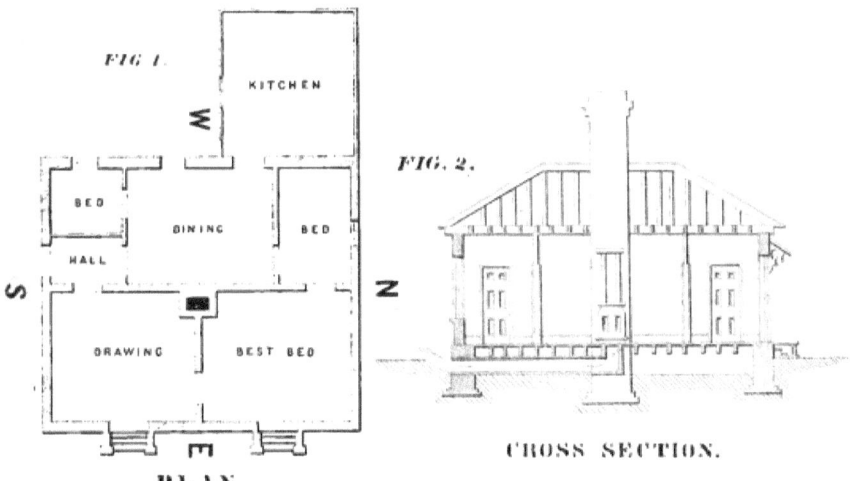

PLAN OF SMALL COTTAGE 29 × 24.
Designed and Drawn by W. Kracme Tamkins Arch.
ST. MARYS, C.W.

CROSS SECTION.

SUPPLIED WITH SHERIFF RUTTANS WARMING & VENTILATING APPARATUS.

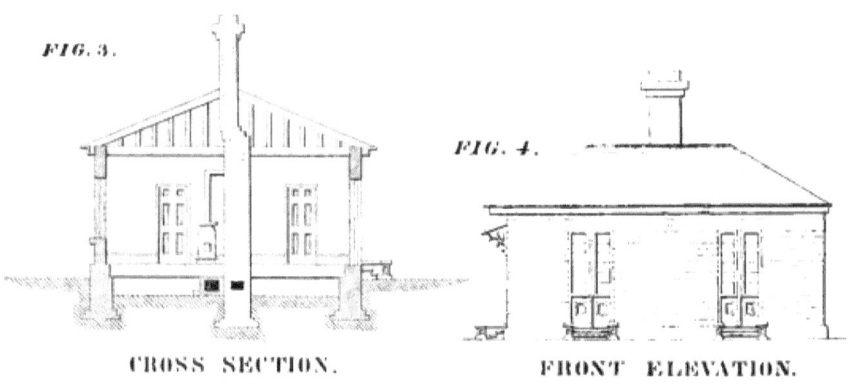

CROSS SECTION.

FRONT ELEVATION.

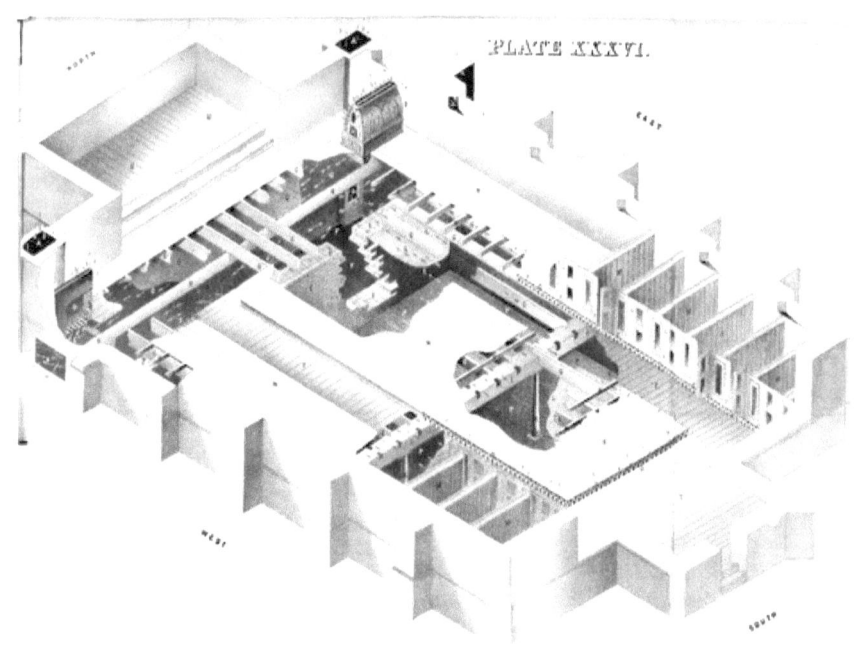

PLATE XXXVI.

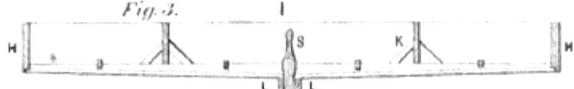

PLATE XXXVII.

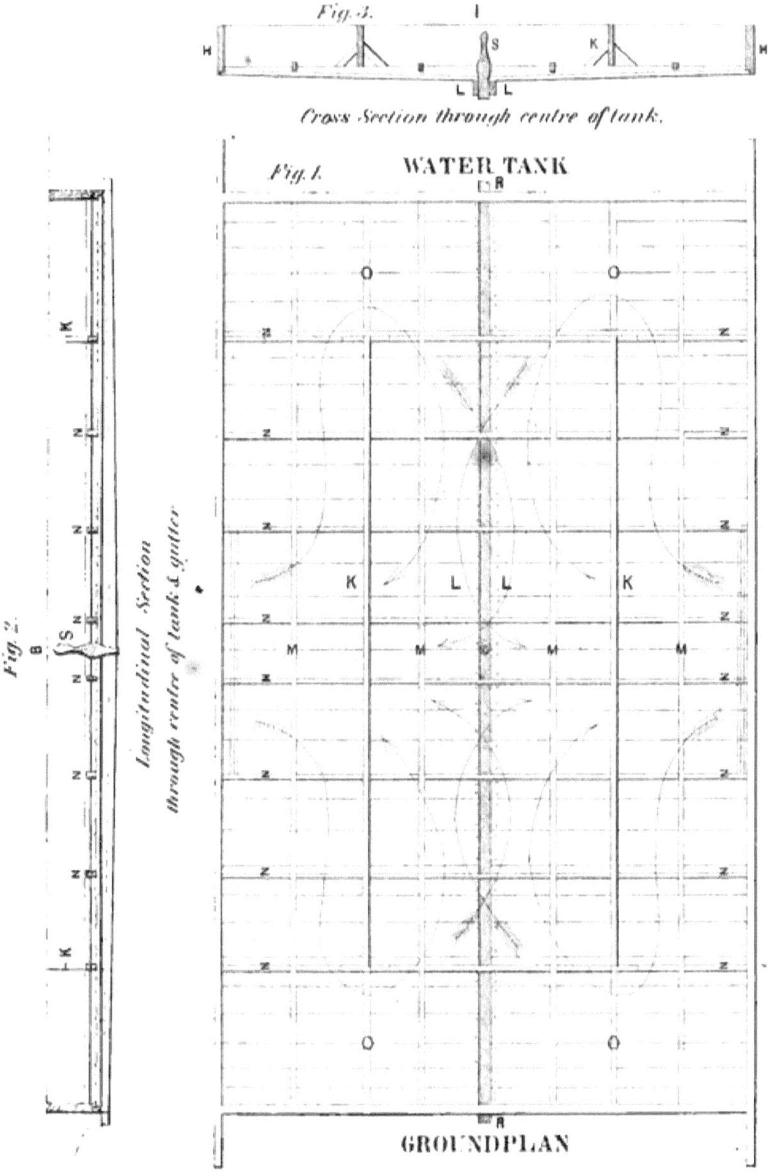

Fig. 3.
I
H
S
K
H
L
L
Cross Section through centre of tank.
Fig. 1.
WATER TANK
R
O
O
K
L
L
K
M
M
M
M
M
O
O
R
GROUNDPLAN
Fig. 2.
K
S
K
Longitudinal Section
through centre of tank & gutter

Scale 3/8 of an inch to the foot.

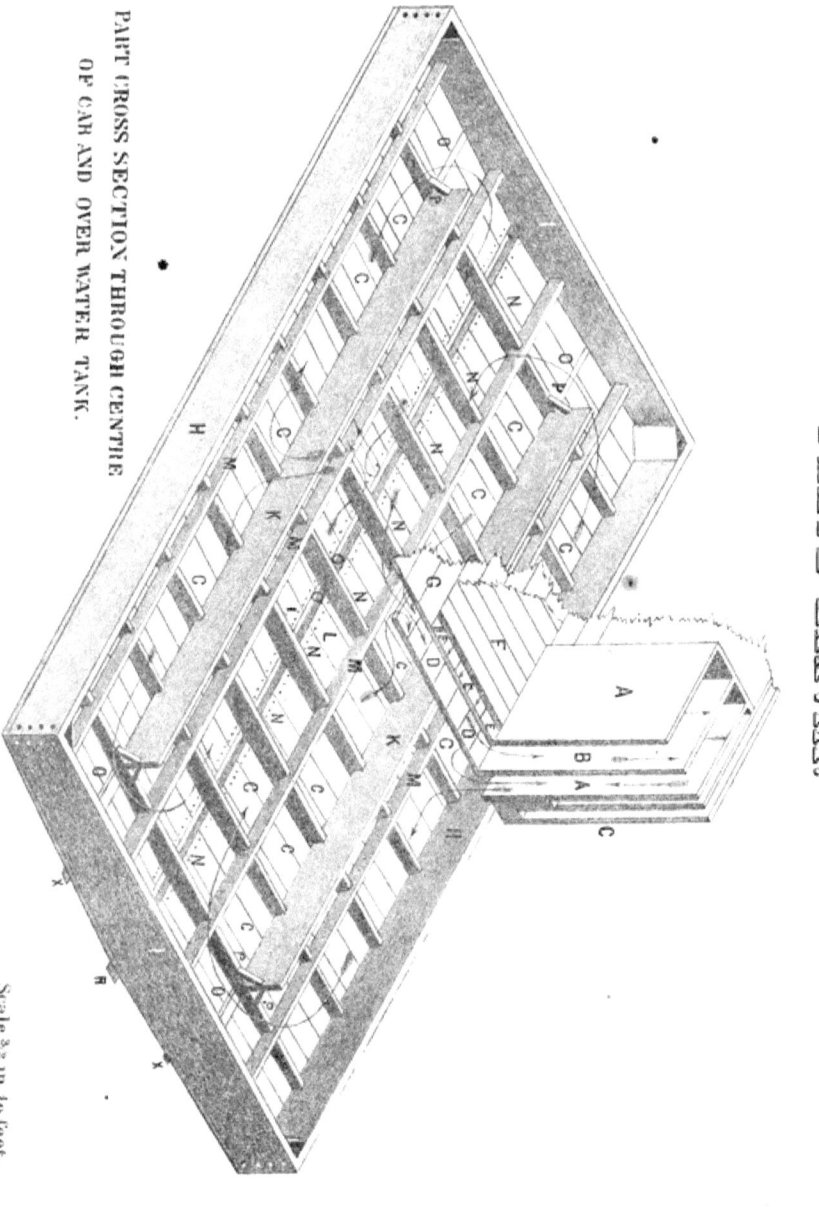

PLATE XXXVIII.

PART CROSS SECTION THROUGH CENTRE OF CAR AND OVER WATER TANK.

Scale 3/8 in. to foot.

PLATE XXXIX

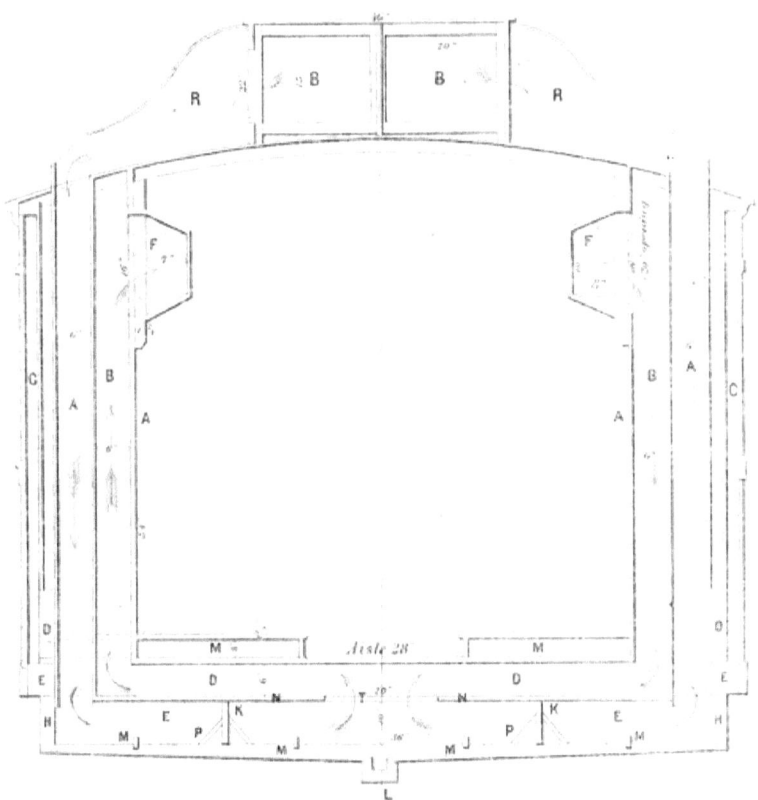

CROSS SECTION

Scale ½ of an inch to the foot

PLATE XL.

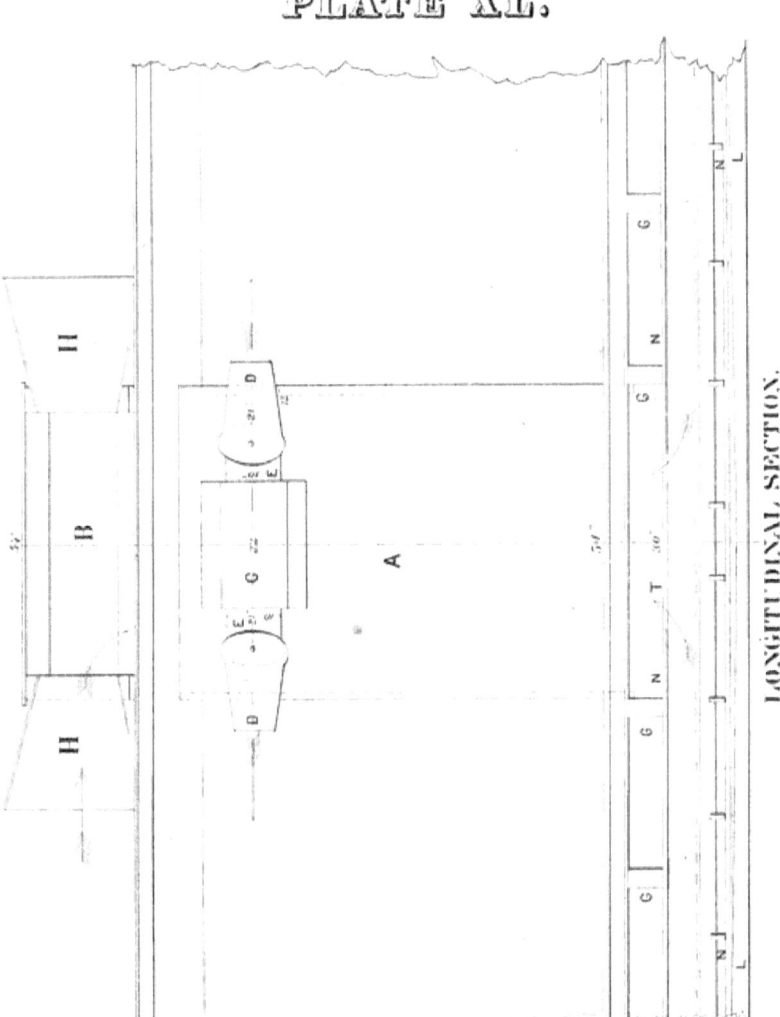

PLATE XLI.

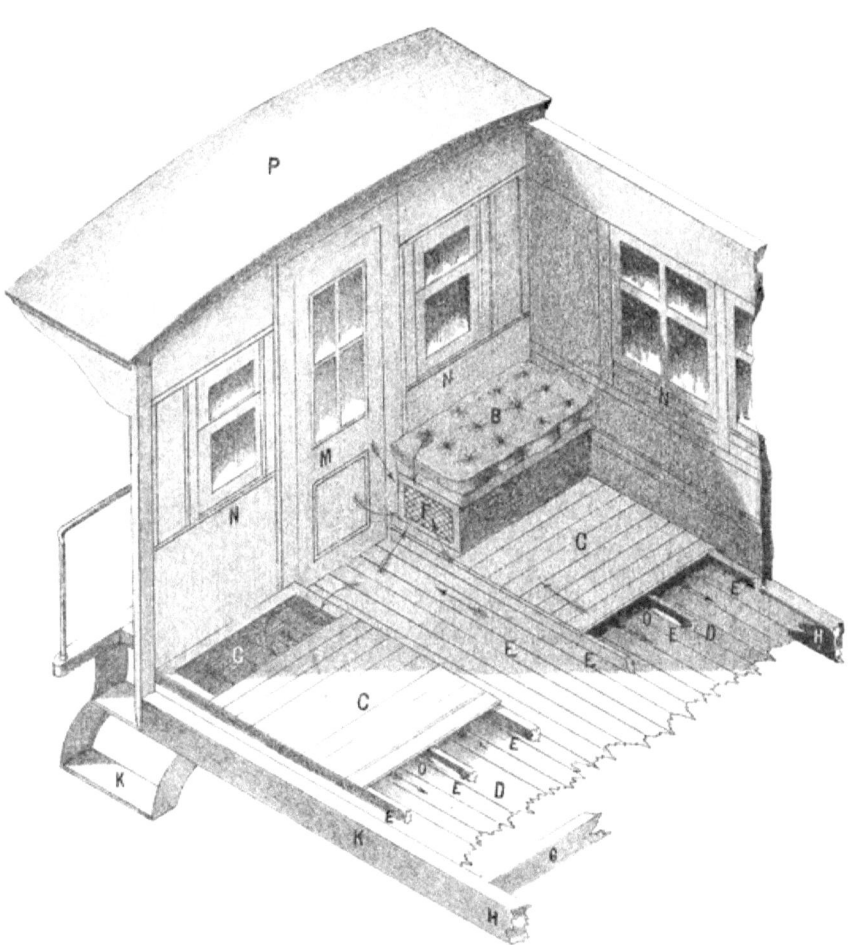

FOUL AIR RECEIVING END.

PLATE XLII.

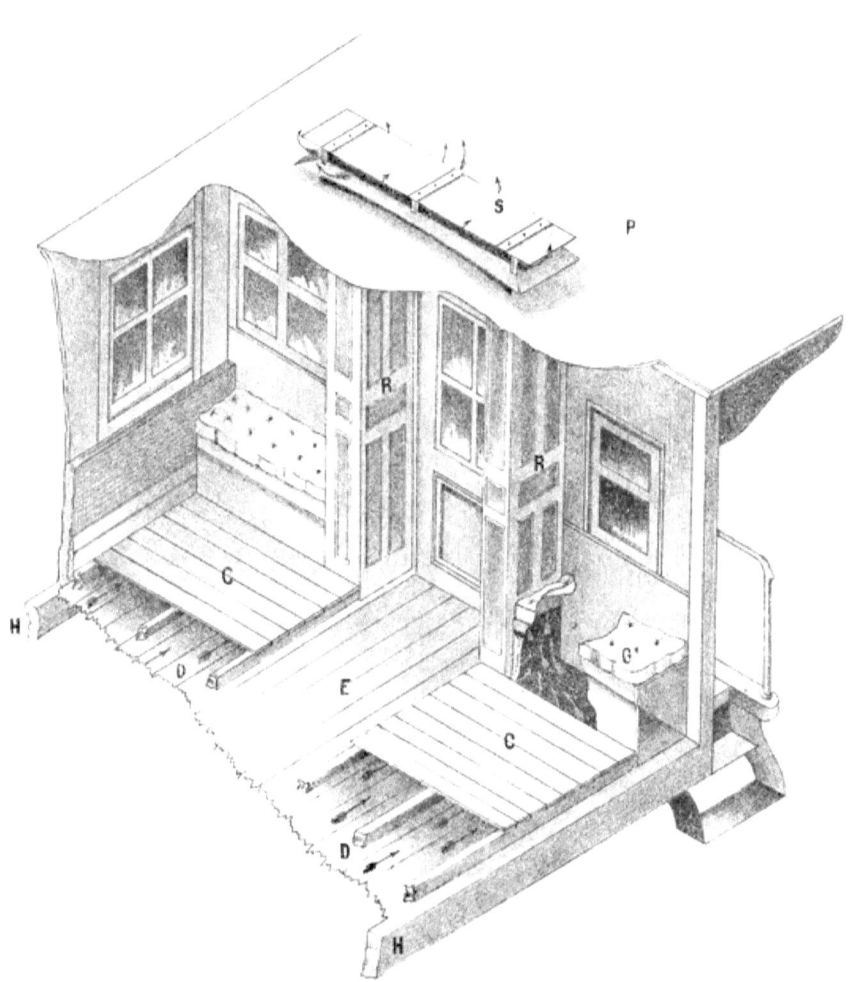

FOUL AIR EXHAUSTING END.

PLATE XLIII

SOMETRICAL VIEW BEING A CROSS AS WELL AS A LONGITUDINAL SECTION OF THE CENTRE PART OF THE CAR.

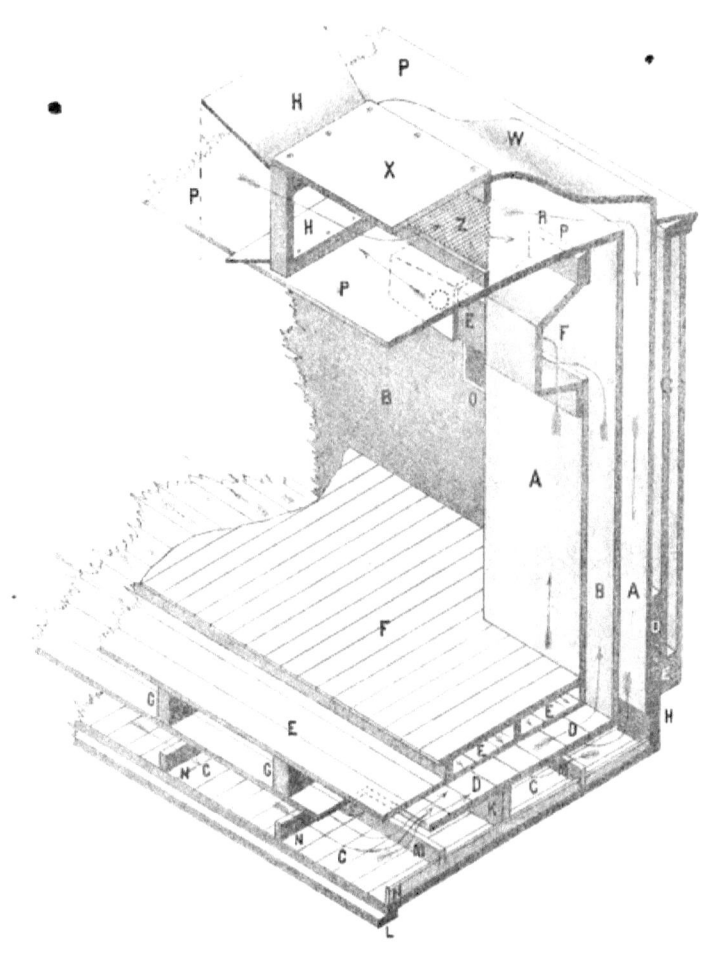

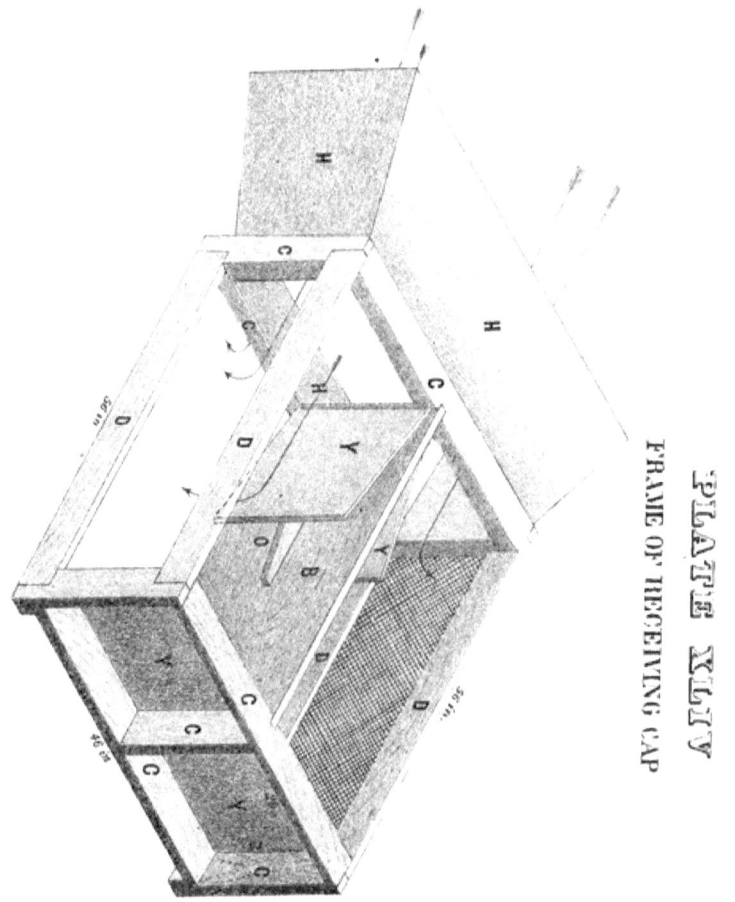

PLATE XLIV
FRAME OF RECEIVING CAP

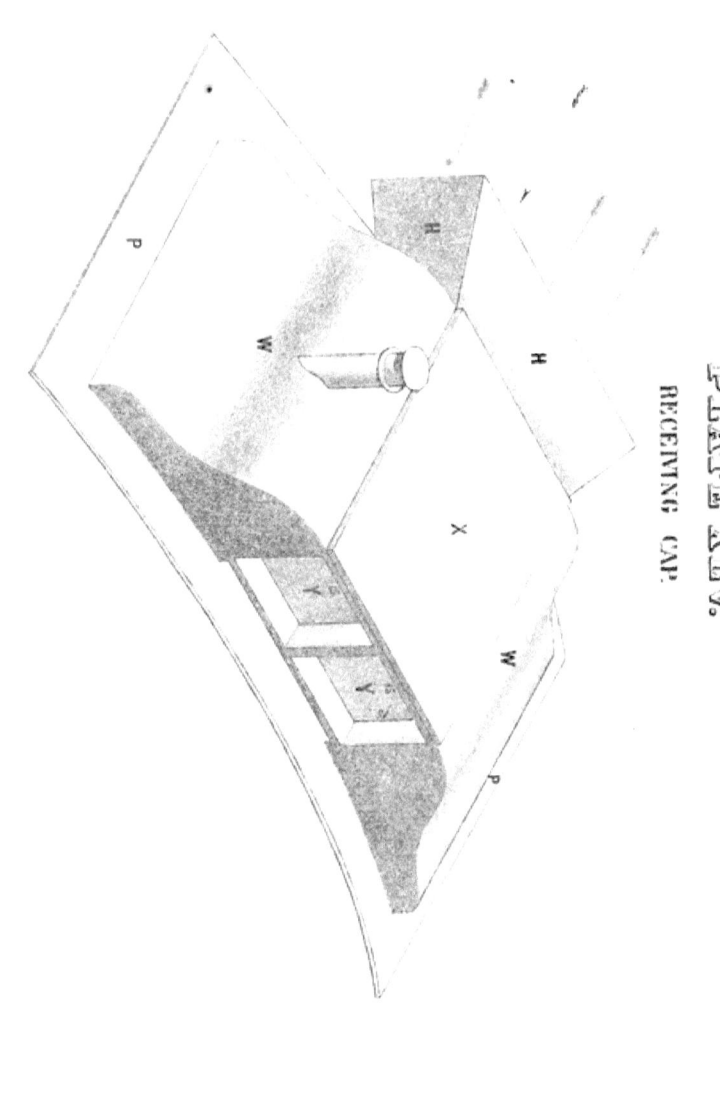

PLATE XLV.
RECEIVING CAP.

PLATE XLVI.

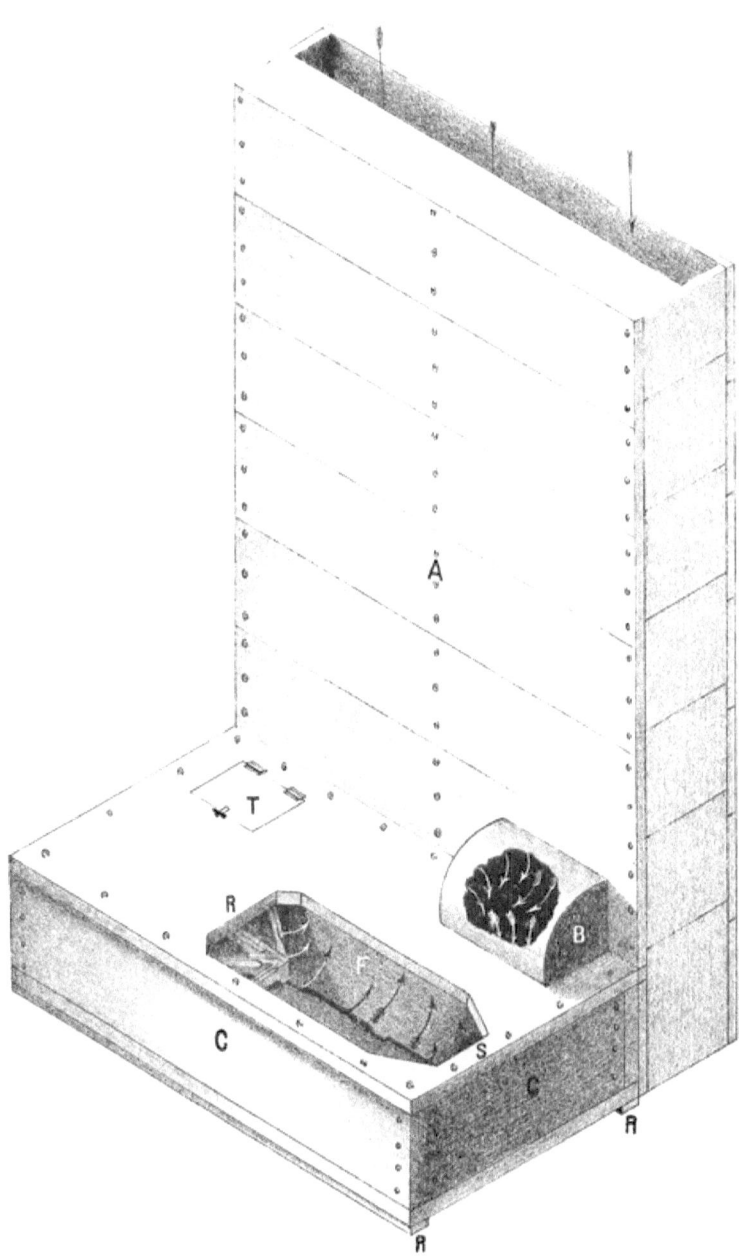

Scale ¾ in to the Foot.

PLATE XLVII.

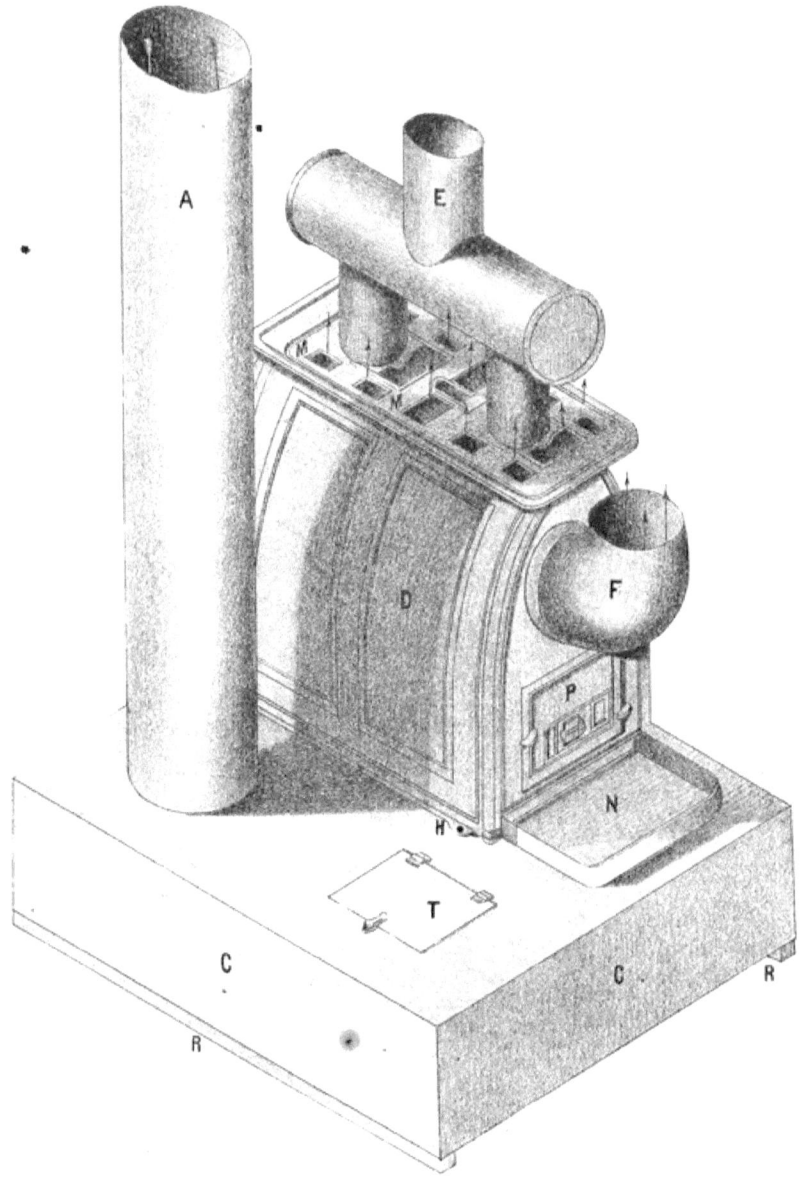

AIR WARMER
OR STOVE

PLATE XLVIII

RECEIVING CAP FOR WINTER WARMING AND VENTILATION OF CARS.

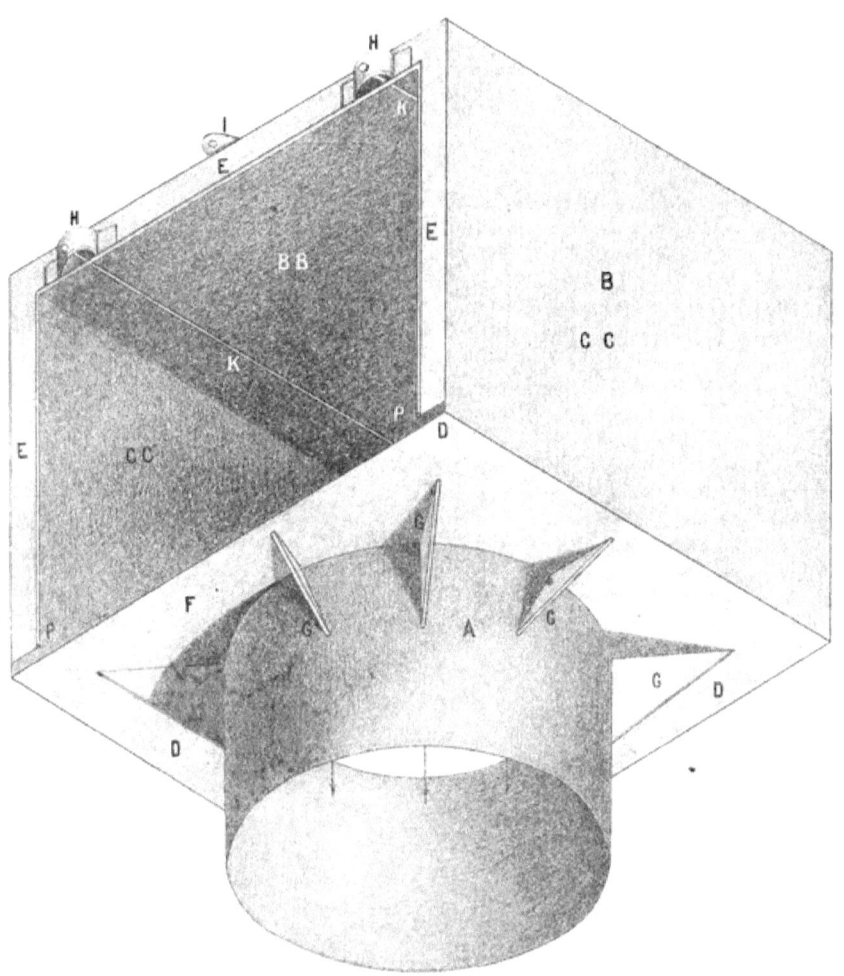

RECEIVING CAP
BOTTOM VIEW.

Scale ¼ full size or 3 in. to the foot

PLATE XLIX.

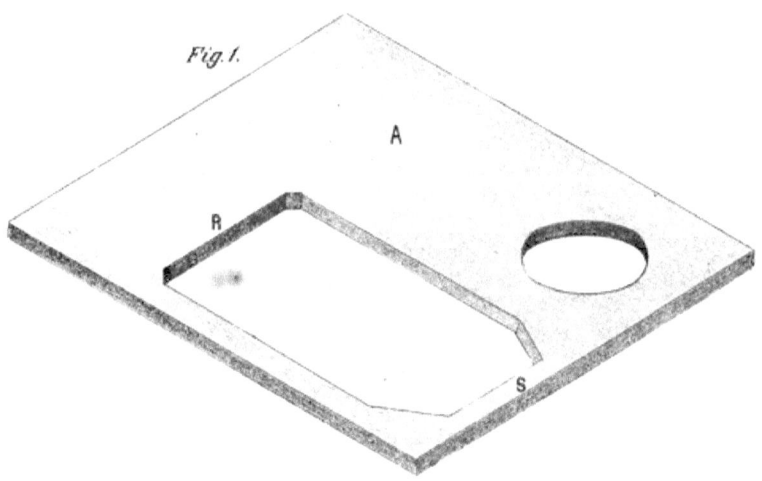

Fig. 1.

R BOX, AND ITS COVER A, UPON WHICH THE AIR WARMER OR STOVE STANDS

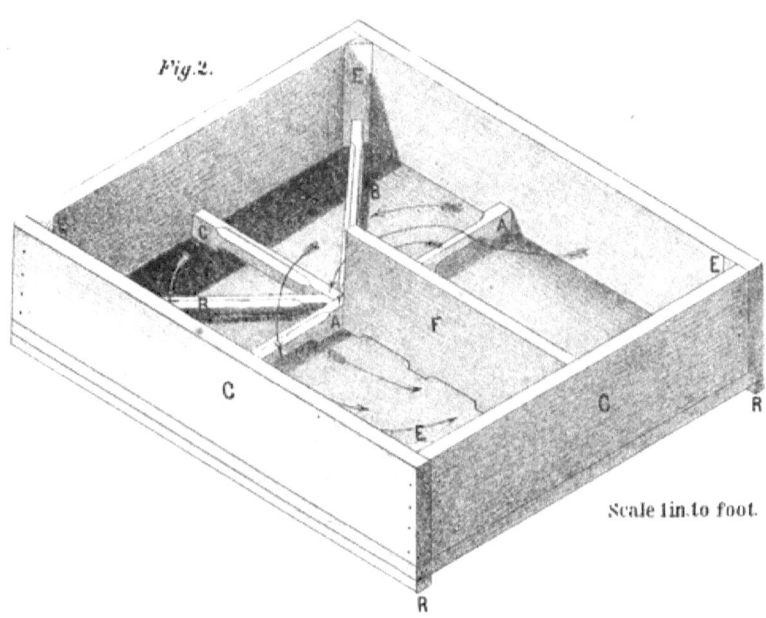

Fig. 2.

Scale 1 in. to foot.

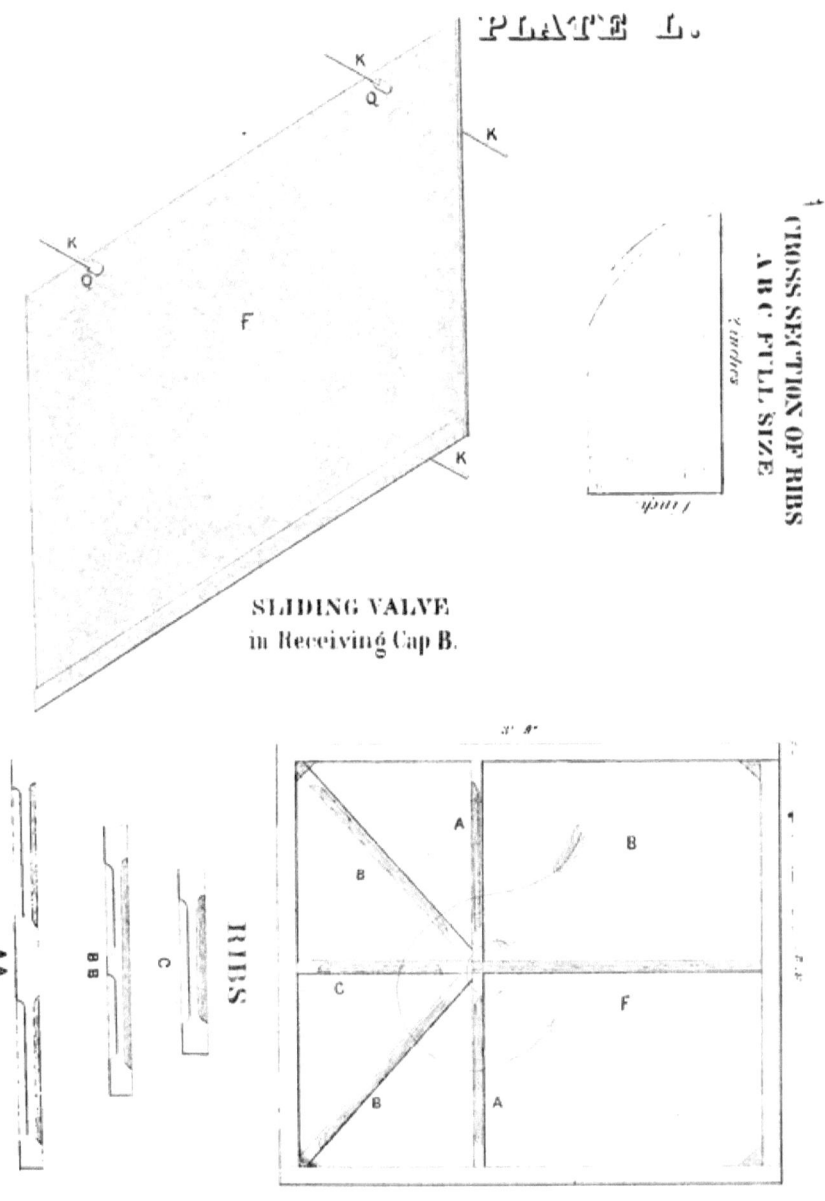

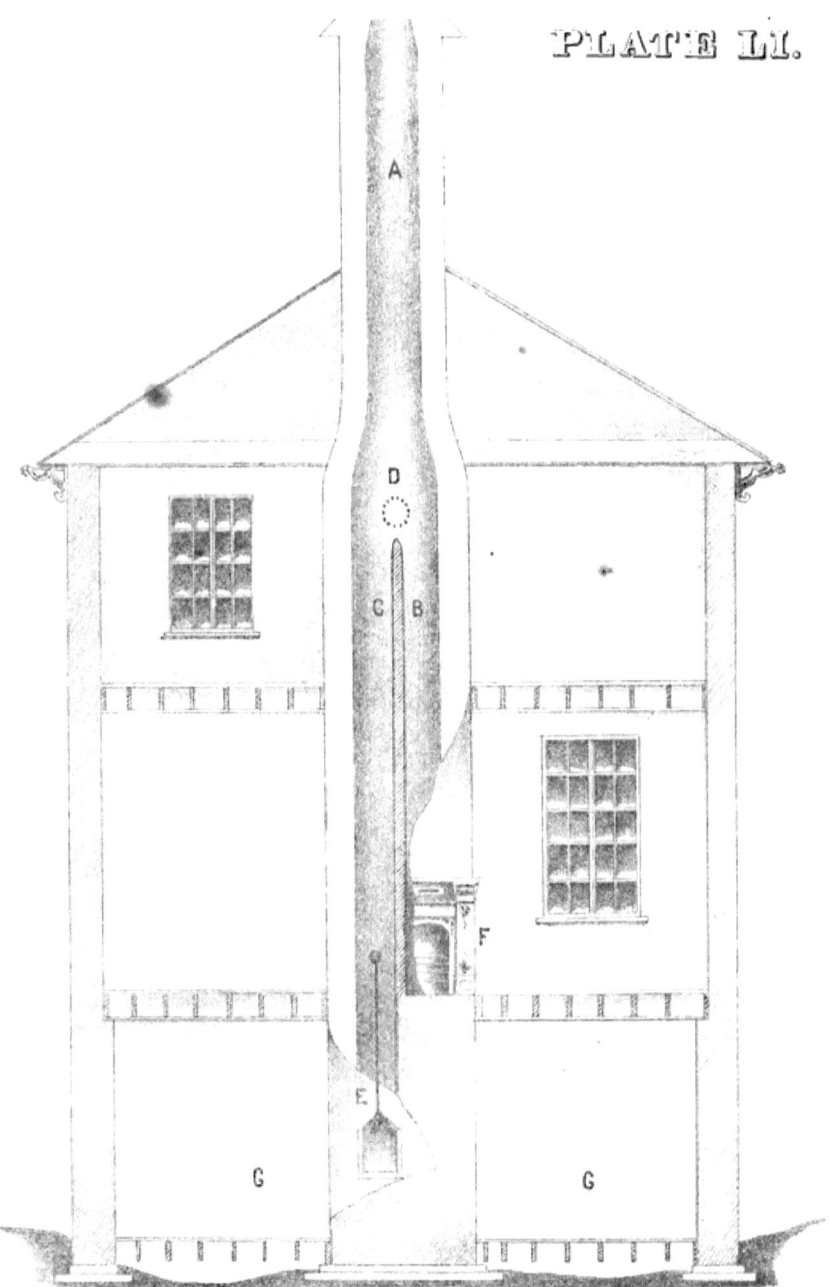

PLATE LI.

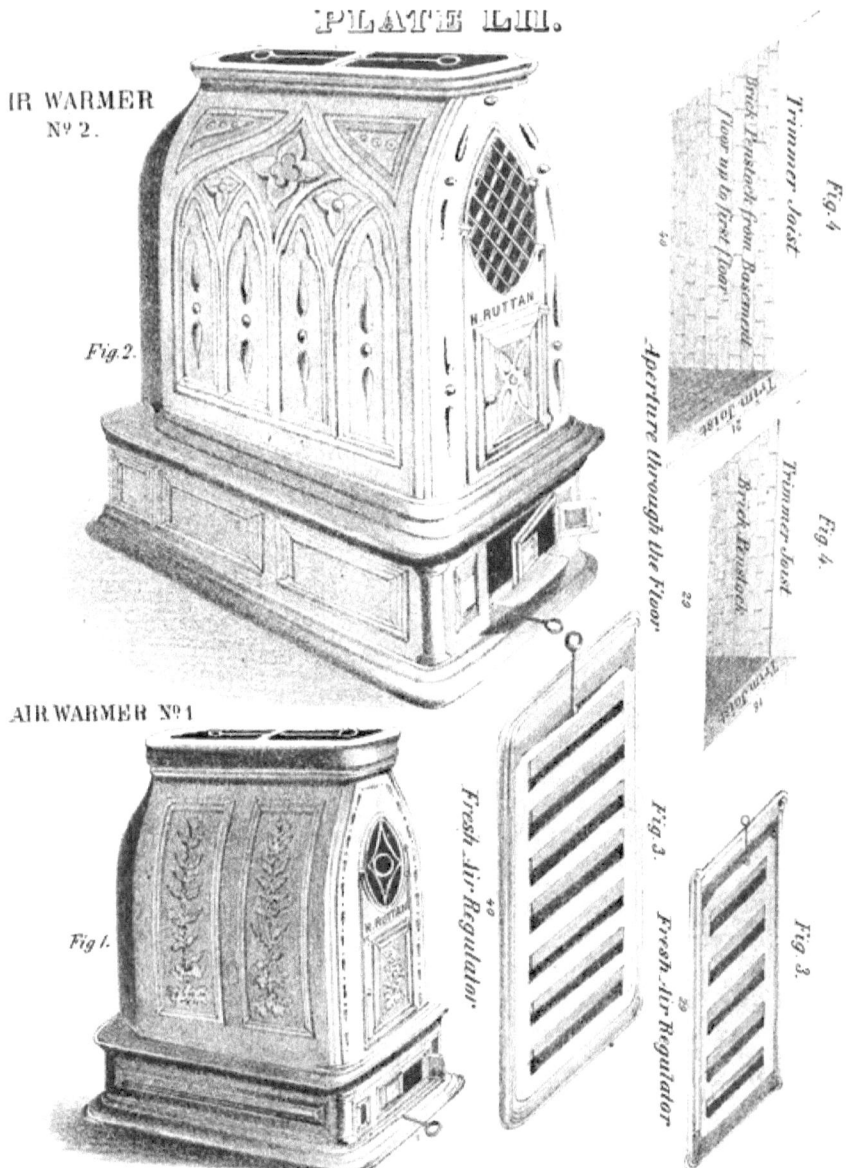

PLATE LII.

AIR WARMER Nº 2.

Fig. 2.

AIR WARMER Nº 1.

Fig. 1.

Fig. 4. Trimmer Joist — Brick Pinstock from Basement floor up to first floor

Fig. 4. Trimmer Joist — Brick Pinstock

Aperture through the Floor

Fresh Air Regulator

Fig. 3. Fresh Air Regulator

Fig. 3.

PLATE LIII.

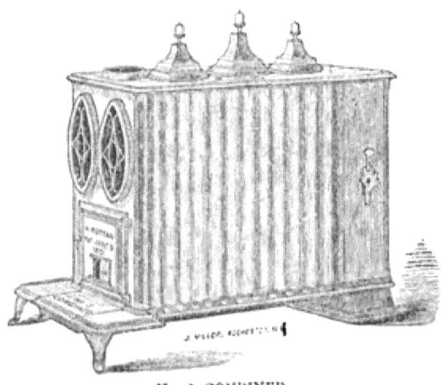

No. 5 COMBINED.

CAR-WARMER.

No. 3 COMBINED.

No. 4.

PLATE LIV.

www.ingramcontent.com/pod-product-compliance
Lightning Source LLC
Chambersburg PA
CBHW031819230426
43669CB00009B/1194